AF213082

Sustainable Civil Infrastructures

Editor-in-chief

Hany Farouk Shehata, Cairo, Egypt

Advisory Board

Khalid M. ElZahaby, Giza, Egypt
Dar Hao Chen, Austin, USA

Steering Editorial Committee

Dar Hao Chen, Texas A&M University, USA
Jia-Ruey Chang, National Ilan University, Taiwan
Hadi Khabbaz, University of Technology Sydney, Australia
Shih-Huang Chen, National Central University, Taiwan
Jinfeng Wang, Zhejiang University, China

About this Series

Sustainable Infrastructure impacts our well-being and day-to-day lives. The infrastructures we are building today will shape our lives tomorrow. The complex and diverse nature of the impacts due to weather extremes on transportation and civil infrastructures can be seen in our roadways, bridges, and buildings. Extreme summer temperatures, droughts, flash floods, and rising numbers of freeze-thaw cycles pose challenges for civil infrastructure and can endanger public safety. We constantly hear how civil infrastructures need constant attention, preservation, and upgrading. Such improvements and developments would obviously benefit from our desired book series that provide sustainable engineering materials and designs. The economic impact is huge and much research has been conducted worldwide. The future holds many opportunities, not only for researchers in a given country, but also for the worldwide field engineers who apply and implement these technologies. We believe that no approach can succeed if it does not unite the efforts of various engineering disciplines from all over the world under one umbrella to offer a beacon of modern solutions to the global infrastructure. Experts from the various engineering disciplines around the globe will participate in this series, including: Geotechnical, Geological, Geoscience, Petroleum, Structural, Transportation, Bridge, Infrastructure, Energy, Architectural, Chemical and Materials, and other related Engineering disciplines.

More information about this series at http://www.springer.com/series/15140

Adam Sevi · Jose Neves
Honghua Zhao
Editors

Enhancements in Applied Geomechanics, Mining, and Excavation Simulation and Analysis

Proceedings of the 5th GeoChina International Conference 2018 – Civil Infrastructures Confronting Severe Weathers and Climate Changes: From Failure to Sustainability, held on July 23 to 25, 2018 in HangZhou, China

Springer

Editors
Adam Sevi
Department of Civil, Environmental, and Construction Engineering
Norwich University
Northfield, VT, USA

Honghua Zhao
Department of Engineering Mechanics
Dalian University of Technology
Dalian, China

Jose Neves
Department of Civil Engineering, Architecture and Georesources, IST
Universidade de Lisboa
Lisbon, Portugal

ISSN 2366-3405 ISSN 2366-3413 (electronic)
Sustainable Civil Infrastructures
ISBN 978-3-319-95644-2 ISBN 978-3-319-95645-9 (eBook)
https://doi.org/10.1007/978-3-319-95645-9

Library of Congress Control Number: 2018948650

© Springer International Publishing AG, part of Springer Nature 2019
This work is subject to copyright. All rights are reserved by the Publisher, whether the whole or part of the material is concerned, specifically the rights of translation, reprinting, reuse of illustrations, recitation, broadcasting, reproduction on microfilms or in any other physical way, and transmission or information storage and retrieval, electronic adaptation, computer software, or by similar or dissimilar methodology now known or hereafter developed.
The use of general descriptive names, registered names, trademarks, service marks, etc. in this publication does not imply, even in the absence of a specific statement, that such names are exempt from the relevant protective laws and regulations and therefore free for general use.
The publisher, the authors and the editors are safe to assume that the advice and information in this book are believed to be true and accurate at the date of publication. Neither the publisher nor the authors or the editors give a warranty, express or implied, with respect to the material contained herein or for any errors or omissions that may have been made. The publisher remains neutral with regard to jurisdictional claims in published maps and institutional affiliations.

This Springer imprint is published by the registered company Springer Nature Switzerland AG
The registered company address is: Gewerbestrasse 11, 6330 Cham, Switzerland

Contents

Introduction

Growing international populations and commerce necessitates increases in infrastructure capacity worldwide. Concurrently, climate change and severe weather events challenge the durability of these structures. This international conference provides a forum for bringing knowledge from practitioners and designers together. This publication of technical papers focusing on geomechanics, mining, and excavation simulation, and analysis is a small contribution to the ongoing challenge of providing sustainable civil infrastructures.

This publication would not have been possible without the efforts of numerous anonymous technical reviewers, working in conjunction with the authors, to shape these technical papers to be most useful to future civil practitioners and designers. Each paper received at least two full reviews, with volume editors supervising this effort. The guidance of the secretary generals, Dr. Jinfeng Wang and Dr. Dar Hao Chen must also be noted for coordinating this volumes contribution to the greater presentation of works contained in the proceedings of this 5th GeoChina International Conference on Civil Infrastructures Confronting Severe Weathers and Climate Changes: From Failure to Sustainability, HangZhou, China, 2018.

Magneto-Gravity Simulation of Cone Penetration Test in Cohesionless Soil Under Small Gravity Fields

Pin-Qiang Mo[1,2(✉)], Feng Gao[2], and Guoqing Zhou[2]

[1] State Key Laboratory for Geomechanics and Deep Underground Engineering, China University of Mining and Technology, Xuzhou 221116, Jiangsu, China
pinqiang.mo@cumt.edu.cn, chen1234@gmail.com

[2] School of Mechanics and Civil Engineering, China University of Mining and Technology, Xuzhou 221116, Jiangsu, China

Abstract. Lunar exploration projects have been re-launched recently by countries including America, Russia, Japan, and China, aiming to develop the resource exploration outside the Earth. The fundamental understanding of the mechanisms related to the small gravity fields and the specific space regolith is the key to further plans for outpost construction and resource utilization, as well as the site investigation. Cone penetration test is believed to serve as an effective in situ tool for site investigation in deep space exploration. The interpretation of CPT data under small gravity fields is investigated in this paper with the focus on the effect of gravity ranging from 1/6 to 4 g. Linear increase of cone tip resistance is observed for a shallow penetration, whereas the resistance increases with g-level under small gravity fields at a certain depth. The normalised penetration resistance is found to decrease exponentially with the g-level, and a relationship between the normalised penetration resistance and g-level is therefore proposed with comparisons of data from DEM simulation. Correlations between soil properties and CPT measurements are thus modified for the interpretation and application of in situ tests in the near-earth space exploration.

1 Introduction

Near-earth space exploration extends from the Moon to the Mars, after the Apollo program. The plans of return to the Moon have been proposed in the recent decade, following the launched lunar missions, including European SMART-1, Chinese Chang'E, U.S. Lunar Reconnaissance Orbiter, and Google Lunar XPRIZE. Studies about the in situ resource utilization (ISRU) have processed the lunar materials for the construction of lunar habitat structures (Meyers and Toutanji 2007; Nakamura and Senior 2008). The colonization of the Moon starts with the lunar outposts, whereas the excavation and the designed underground colonies require the understanding of geotechnical properties of the lunar regolith.

Cone penetration test (CPT) acts as an important in situ testing tool for the site characterization, owing to its economic advantages and the continuous measurements. However, in the field of aerospace, the gravitational environment on a planetary body varies with its mass; e.g. 1 g indicates the Earth gravity and 1/6 g represents the lunar

© Springer International Publishing AG, part of Springer Nature 2019
A. Sevi et al. (eds.), *Enhancements in Applied Geomechanics, Mining, and Excavation Simulation and Analysis,* Sustainable Civil Infrastructures,
https://doi.org/10.1007/978-3-319-95645-9_1

gravity. In situ testing is important and rather essential for the further space exploration, without taking numerous samples back to earth for investigation. Therefore the interpretation of CPT data is key to obtain the local soil stratigraphy and soil properties. DEM simulation has been conducted to investigate the effects of gravity on the correlations between CPT measurements and soil properties (e.g. Jiang and Wang 2013), while the experimental study is still rather limited.

Physical modelling of geotechnical problems under small gravity fields, especially for gravity smaller than the Earth gravity, is of interest to space and geotechnical engineers. The parabolic flight aircraft, microgravity rocket or the drop tower were used to create the microgravity environments, while they were not widely applied due to their extremely high cost and the limited testing duration (Johnson et al. 1970; Sture et al. 1998; Colwell and Taylor 1999; Thomas et al. 2000). Although the tilting method (Tateyama 2007; Zou et al. 2015) and the hydraulic gradient similitude method (Zelikson 1969; Dou and Byrne 1996) were adopted along with the additional body force or water pressure, a reliable and economical testing method for small gravity fields is still desired for studies of geotechnical problems.

This paper aims to provide results of the preliminary tests of CPT in cohesionless soil under small gravity fields. The magneto-gravity modelling equipment is adopted to simulate the small gravity fields with a magnetic acceleration against the earth gravity. The cohesionless soil is used in the magneto-gravity fields to represent the earth, lunar and martian regolith. The effect of gravity on the penetration resistance in cohesionless soil is investigated by varying g-level from 1/6 to 4 g. Correlations between CPT data and soil properties are then examined to provide insights into the penetration mechanisms of CPT under small gravity fields, which could contribute to the in situ resource utilization in the near-earth space exploration.

2 Testing Methodology

2.1 Simulation of Small Gravity Fields

The experimental tests of CPT are carried out using a testing apparatus for the simulation of small gravity fields, and the method is also termed as magneto-gravity simulation. The geotechnical magneto-gravity model testing equipment was designed based at China University of Mining and Technology, which has the ability to simulate gravity fields from 0 to 6 g (g indicates the earth gravity, $g = 9.81\ m/s^2$). This equipment consists of testing container, power supply system, cooling system, and the data acquisition system, as shown in Fig. 1. The testing container includes copper coils for the generation of gradient magnetic field, in cooperation with the power supply and cooling system. The details on the theory of magnetic similitude gravity and development of the testing equipment are addressed in the unpublished report by Zhou et al. (2016).

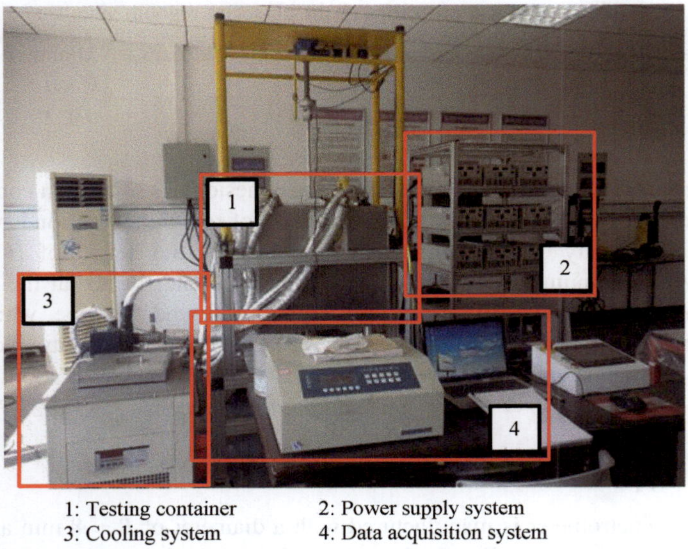

Fig. 1. The geotechnical magneto-gravity model testing equipment

2.2 Planetary Regolith Simulant

Planetary regolith is simulated by a type of magnetically sensitive granular material, following the development of Li et al. (2012). The regolith simulant, made from equal portions of cement (type: 62.5R) and Fe3Q4 powder (type: BMMF-1), is formed by breaking the cured cubes (cement-water ratio is 2.5:1, curing period is 28 days) into magnetic particles and is taken as the cohesionless soil with a reasonable particle distribution, as can be seen in Fig. 2a. The basic physical properties of the planetary regolith simulant is provided in Table 1.

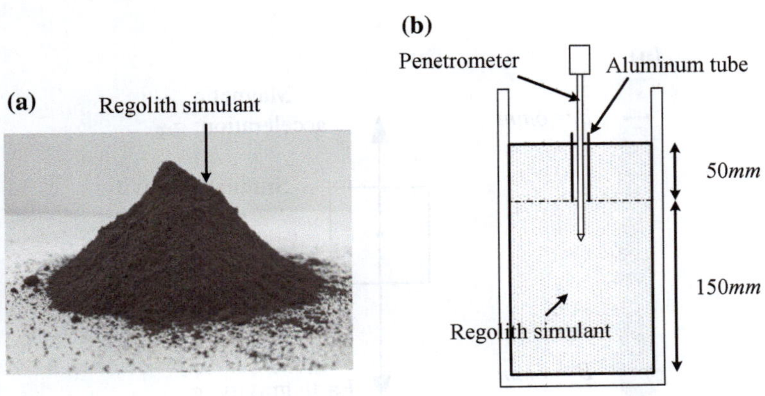

Fig. 2. The planetary regolith simulant and sample preparation

Table 1. Physical properties of the planetary regolith simulant

Specific gravity (G_s)	Mean particle size (d_{50}/mm)	Maximum void ratio (e_{max})	Minimum void ratio (e_{min})
2.930	0.34	1.704	0.032

Testing samples are prepared by pouring the cohesionless soil into a container with inner diameter of $D = 86$ mm. Every 25 mm height of soil with vibration is validated to ensure the uniformity of the samples, until it reaches the height of $H = 150$ mm (Fig. 2b). An aluminum tube is then placed on the sample surface at the location of penetration, and additional 50 mm of soil is poured around the tube, which aims to provide surcharge avoiding the smear of soil surface caused by the attraction between particles in the magnetic field. The void ratio of the prepared samples is about $e = 1.325$, that leads to the density $\rho = 1.26\ \mathrm{g/cm^3}$ and the relative density $D_r = 22.7\%$.

2.3 CPT Apparatus

A miniature penetrometer is manufactured with a diameter of $B = 8$ mm and an apex angle of 60°, as presented in Fig. 3a. Noted that the standard cone size for CPT is about 35.7 mm, which is 4.5 times the mini-penetrometer; therefore the scale effect with factor of 4.5 should be considered in this physical modelling. The cone diameter to particle size ratio $B/d_{50} = 23.5$ is greater than 20, which is the limit value to avoid the particle size effects reported by Gui et al. (1998). On the other hand, the container to probe diameter ratio $D/B = 10.8$ is used to reduce the effect of side wall. The design of shrunken shaft ($b = 6$ mm) aims to allow the relaxation of soil after the probe shoulder and to reduce the fraction of shaft friction. In this way, the measured penetration load is taken as the cone tip load for shallow penetration, neglecting the amount of friction. A loading frame is mounted above the testing container, as shown in Fig. 1, allowing the penetration at a speed of 0.55 mm/s with the installed actuator. An 'S' type load cell is used for the record of penetration load with a frequency of 1 Hz, and the total penetration depth is approximate 130 mm, avoiding the effect of bottom base.

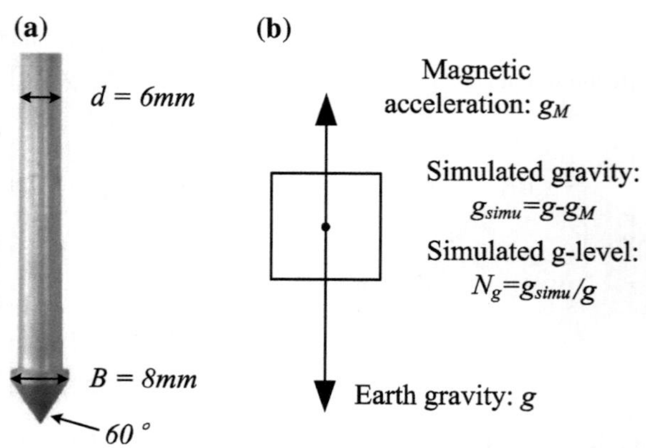

Fig. 3. Schematic of **a** miniature penetrometer; **b** g-level

The CPT tests are conducted under different gravity fields, including 1/6 g (the lunar gravity), 1/3, 2/3, 1, 2, 4 g. The g-level N_g, defined as:

$$N_g = \frac{g - g_M}{g} = \frac{g_{simu}}{g} \tag{1}$$

where g_M is the magnetic acceleration, is simulated by the combination of earth gravity and the magnetic acceleration, as illustrated in Fig. 3b. Note that repeatability tests are also conducted to maintain the reliability of the experimental data.

3 Results and Discussion

3.1 Results of Cone Penetration Tests

The cone tip resistance q_c is obtained by dividing the recorded penetration force with the cross-sectional area of the probe during its insertion. At least three repeatability tests were conducted for each test with a given g-level, and the average value of penetration resistance is thus calculated for further investigation. The testing data for $N_g = 1/6$ is shown in Fig. 4a, as well as the averaging cone tip resistance against the penetration depth. The initial rapid increase of q_c in the first 8 mm ($z_1 = B$) is attributed to the upper 50 mm of surcharge, which was applied to avoid the smeared surface by the magnetization. After that, a nearly linear growth of q_c is observed for shallow penetration, in consistence with the centrifuge tests (Gui et al. 1998; Xu 2007; Mo et al. 2015). For a relatively deep penetration ($z > 15B$), the growth tends to be faster, presumably owing to the bottom effect and the gained shaft friction of the miniature probe.

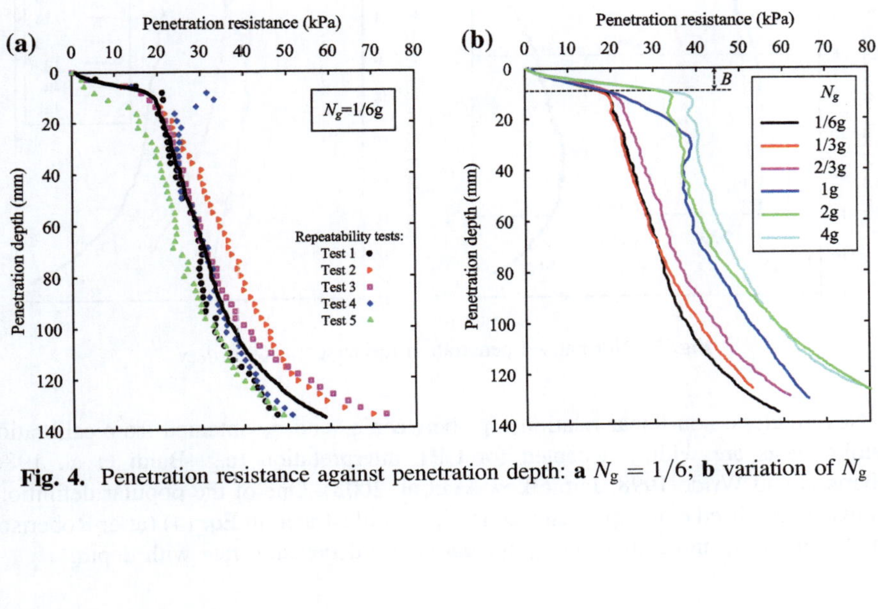

Fig. 4. Penetration resistance against penetration depth: **a** $N_g = 1/6$; **b** variation of N_g

The effect of gravity is presented in Fig. 4b, with the variation of g-level N_g ranging from 1/6 to 4. Linear increase of the cone tip resistance is again observed for a shallow penetration. It is obvious that q_c increases with g-level under small gravity fields at a certain depth, where the in situ stress is higher for a larger g-level. However, the increase seems to be nonlinear, and the effect of gravity is more significant for g-level smaller than 1 g.

In view of the observed linear relationship between the cone tip resistance and the depth (i.e. the vertical effective stress), it is reasonable to apply the normalization of CPT data following Bolton et al. (1993), as defined in Eqs. (2) and (3). This normalization is also consistent with the definition of cone factor in sand for bearing capacity analysis, where $N_q = q_c/\sigma_v = Q + 1$.

$$Q = \frac{q_c - \sigma_v}{\sigma_v} \tag{2}$$

$$Z = \frac{z}{B} \tag{3}$$

The results of normalised penetration resistance are provided in Fig. 5a. Note that the initial vertical stress is a combination from both the soil gravity and the surcharge. A comprehensive constant value of Q is obtained for the normalised penetration depth between 6 and 14. Accordingly, Q decreases with the g-level, as also reported by Bolton et al. (1999) with series of centrifuge tests.

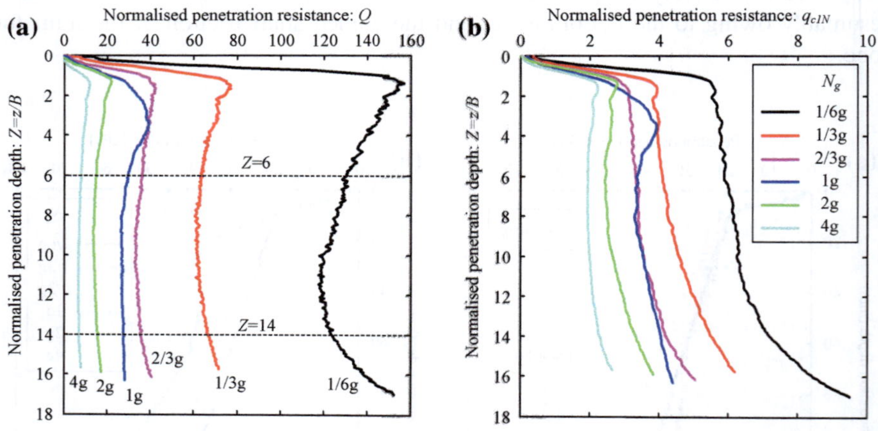

Fig. 5. Normalised penetration resistance: **a** Q; **b** q_{c1N}

Nevertheless, non-linear relationships between q_c and σ_v obtained from calibration chamber tests are widely accepted for CPT interpretation (e.g. Baldi et al. 1986; Robertson and Wride 1998; Jamiolkowski et al. 2003). One of the popular definitions of stress-normalised cone tip resistance is q_{c1N}, as illustrated in Eq. (4) (after Robertson and Wride 1998), indicating that q_c increases at a decreasing rate with depth.

$$q_{c1N} = \frac{(q_c - \sigma_v)/\sigma_0}{(\sigma_v/\sigma_0)^{0.5}} \quad (4)$$

where σ_0 is the reference stress = 100 kPa. The corresponding results are shown in Fig. 5b. Although the decrease of q_{c1N} is also found, it can be concluded that Q is a more appropriate normalization to achieve a constant value for penetration under a small gravity field.

3.2 Normalised Penetration Resistance Against g-Level

In terms of the decrease of the normalised penetration resistance Q with g-level, the observed data within the range $6 < Z < 14$ is plotted in Fig. 6a, showing an exponential decrease. The variation of Q also fades away gradually with the increasing g-level. When the results of Q is normalised by the magnitude of 1 g test Q_{1g}, the exponential decrease can be predicted as the following expression, as shown in Fig. 6b.

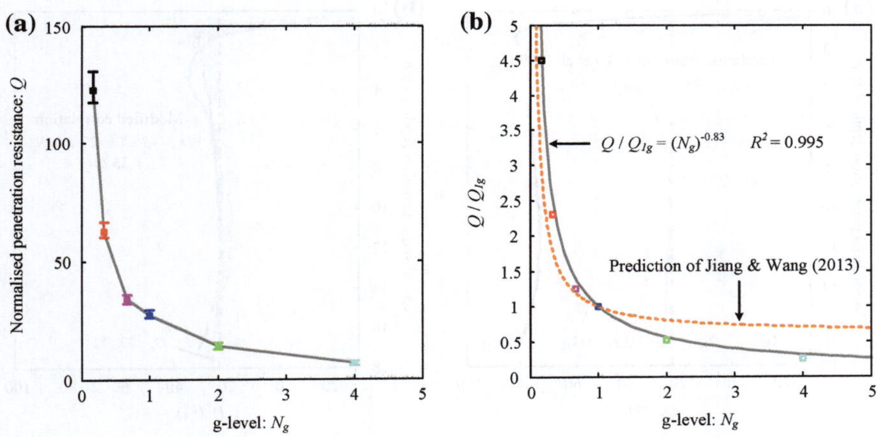

Fig. 6. Normalised penetration resistance against g-level: **a** Q; **b** Q/Q_{1g}

$$Q/Q_{1g} = N_g^{-0.83} \quad (5)$$

Note that this expression is obtained based on the loose sand tests for small gravity fields ($N_g < 5$), and comprehensive verification is suggested when it is applied for larger magnitude of g-level. The results are also compared with the prediction of Jiang and Wang (2013), which was deduced from two-dimensional DEM simulation of CPT under different gravity fields. The dramatic decrease of Q/Q_{1g} for $N_g < 1$ is comparative to the experimental data in this paper, whereas the prediction of Q was much higher for DEM simulation of large g-level tests.

3.3 Back Analysis of Relative Density

Relative density D_r is an important indication to the state of soil and for the prediction of mechanical properties. Many correlations were also proposed to relate CPT data to the relative density based on calibration chamber tests (e.g. Lancellotta 1983; Tatsuoka et al. 1990; Jamiolkowski et al. 2003; Bolton and Gui 1993). However, the empirical correlations are only suitable for penetration under earth gravity with relatively high stress conditions, and the direct application would introduce exaggerated false to the prediction. As illustrated in Fig. 7a, the correlation of Tatsuoka et al. (1990) is used for the back analysis of relative density. The sample preparation shows an average D_r of 22.7%, while the predictions vary with significant ranges, even with a negative value for the 4 g tests. It is worth noting that the prediction of D_r for the 1 g tests is close to the measurement before CPT tests. Therefore, the correlation for penetration under different gravity is modified based on the 1 g correlation and the prediction of normalised penetration resistance in Eq. (5), as shown in the following expression.

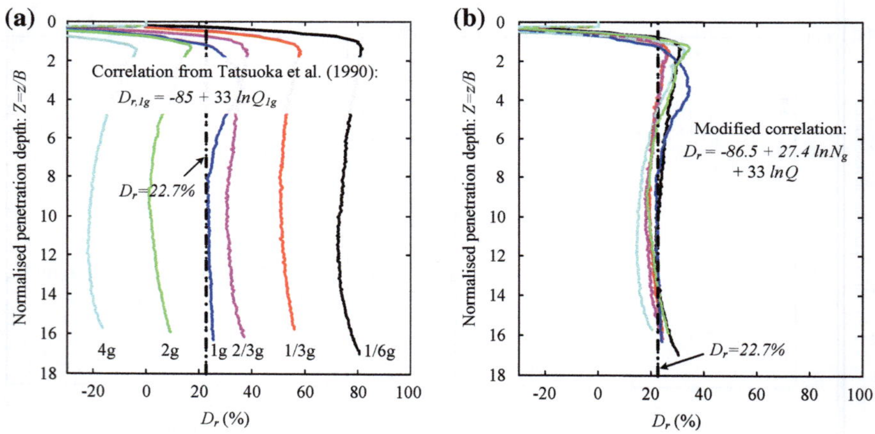

Fig. 7. Back analysis of relative density: **a** correlation of Tatsuoka et al. (1990); **b** modified correlation

$$D_{\mathrm{r}} = -86.5 + 27.4 \cdot \ln N_{\mathrm{g}} + 33 \cdot \ln Q \tag{6}$$

The modified results for the back analysis of D_r based on the cone tip resistance are thus found with improved predictions, in consistence with the measured data for different gravity fields (see Fig. 7b).

4 Conclusions

This paper presents the results of preliminary tests of cone penetration testing in cohesionless soil under small gravity fields ranging from 1/6 to 4 g. The gravity varies to simulate the gravitational environment on a planetary body; e.g. 1 g indicates the Earth gravity and 1/6 g represents the Moon gravity. The effects of gravity on the correlations between CPT data and soil properties are investigated by applying a magnetic acceleration against the earth gravity using the magneto-gravity modelling system. Accordingly, the cohesionless soil is made from the cement powder and Fe_3O_4 magnetic powder with a rational particle distribution, simulating the planetary regolith. A miniature penetrometer with diameter of 8 mm is manufactured along with the actuator proving a penetration speed at around 0.55 mm/s.

From the results of penetration resistance, linear increase of q_c is observed for a shallow penetration. It is obvious that the cone tip resistance increases with g-level under small gravity fields at a certain depth, where the in situ stress is higher for a larger g-level. Q is found to be a more appropriate normalization to achieve a constant value for penetration under a small gravity field. Additionally, the normalised penetration resistance decreases exponentially with the g-level, and a relationship between Q/Q_{1g} and N_g has been proposed with comparison of DEM simulation. The empirical correlation between CPT measurements and relative density has been modified to include the effects of gravity, and the improved predictions present relatively consistent results with the measured data for different gravity fields. This study examined the influence of planetary gravity to the interpretation of in situ tests, which would contribute to the understanding of the stratigraphy and properties of the planetary regolith for further in situ resource utilization and the construction of space outposts.

Acknowledgements. The author would like to acknowledge financial support from the National Natural Science Foundation of China (no. 51323004), 111 Project (B14021) and Natural Science Foundation of Jiangsu Province (no. BK20170279).

References

Baldi, G., Bellotti, R., Ghionna, V.N., Jamiolkowski, M., Pasqualini, E.: Interpretation of CPTs and CPTUs; 2nd part: drained penetration of sand. In: Proceedings of the 4th International Geotechnical Seminar, Singapore, pp. 143–156 (1986)

Bolton, M.D., Gui, M.W.: The study of relative density and boundary effects for cone penetration tests in centrifuge. Technical Report CUED/DSOILS/TR256, University of Cambridge (1993)

Bolton, M.D., Gui, M.W., Phillips, R.: Review of miniature soil probes for model tests. In: Proceedings of 11th South East Asia Geotechnical Conference, pp. 85–91 (1993)

Bolton, M.D., Gui, M.W., Garnier, J., Corte, J.F., Bagge, G., Laue, J., Renzi, R.: Centrifuge cone penetration tests in sand. Géotechnique **49**(4), 543–552 (1999)

Colwell, J.E., Taylor, M.: Low-velocity microgravity impact experiments into simulated regolith. Icarus **138**(2), 241–248 (1999)

Dou, H., Byrne, P.M.: Dynamic response of single piles and soil pile interaction. Can. Geotech. J. **33**(1), 80–96 (1996)

Gui, M., Bolton, M.D., Garnier, J., Corte, J.F., Bagge, G., Laue, J., Renzi, R.: Guidelines for cone penetration tests in sand. In: International Conference on Centrifuge Modelling (Centrifuge'98), Balkema, vol. 1, pp. 155–160 (1998)

Jamiolkowski, M., Lo Presti, D.C.F., Manassero, M.: Evaluation of relative density and shear strength of sands from CPT and DMT. Soil Behav. Soft Ground Constr. **7**(119), 201–238 (2003)

Jiang, M.J., Wang, X.X.: Numerical analysis of cone penetration tests under different gravity fields by distinct element method. Rock Soil Mech. **34**(3), 863–873 (2013)

Johnson, S.W., Pyrz, A.P., Lee, D.G.: Simulating the effect of gravitational field and atmosphere on behavior of granular media. J. Spacecr. Rocket **7**(11), 1311–1317 (1970)

Lancellotta, R.: Analisi di affidabilita in ingegneria geotechnica. Technical Report 625, Politernico di Torino (1983)

Li, R.L., Zhou, G.Q., Chen, G.Z., et al.: Development of a new geo-mechanical magnetic sensing soils similar material. Funct. Mater. **19**(43), 2620–2623 (2012)

Meyers, C., Toutanji, H.: Analysis of lunar-habitat structure using waterless concrete and tension glass fibers. J. Aerosp. Eng. **20**(4), 220–226 (2007)

Mo, P.Q., Marshall, A.M., Yu, H.S.: Centrifuge modelling of cone penetration tests in layered soils. Géotechnique **65**(6), 468–481 (2015)

Nakamura, T., Senior, C.L.: Solar thermal power for lunar materials processing. J. Aerosp. Eng. **21**(2), 91–101 (2008)

Robertson, P.K., Wride, C.E.: Evaluating cyclic liquefaction potential using the cone penetration test. Can. Geotech. J. **35**(3), 442–459 (1998)

Sture, S., Costes, N.C., Batiste, S.N., et al.: Mechanics of granular materials at low effective stresses. J. Aerospace Eng. **11**(3), 67–72 (1998)

Tatsuoka, F., Zhou, S., Sato, T., Shibuya, S.: Evaluation method of liquefaction potential and its application. Technical report, Ministry of Education of Japan (1990)

Thomas, V.A., Prasad, N.S., Reddy, C.A.M.: Microgravity research platforms—a study. Cur. Sci. **79**(3), 336–340 (2000)

Xu, X.T.: Investigation of the end bearing performance of displacement piles in sand. Ph.D. thesis, The University of Western Australia (2007)

Zelikson, A.: Geotechnical models using the hydraulic gradient similarity method. Géotechnique **19**(4), 495–508 (1969)

Zhou, G.Q., Li, R.L., Chen, G.Z., Zhao, X.D., Wang, J.Z., Liang, H.C., Lin, C., Lai, Z.J., Chen, X.: Geotechnical magnetic-similitude-gravity model testing method. Unpublished report (2016)

The Effect of Several Parameters on the Behavior of Asphalt Mixture in Libya

Khlifa El Atrash(✉) and Gabriel J. Assaf

École de Technologie Supérieure, Université de Québec, Quebec City, Canada
k_atrch@yahoo.com

Abstract. In Libya, many steps have been taken to extend the life of the pavement and improve the design of asphalt mixture to have a better connection to Libyan climate. This study is to evaluate the effect of some parameters on the performance of the asphalt mixture in hot arid weather as Libya. Two asphalt concrete mixtures were designed using two different binders, Performance Grade binder (PG70-10) and Pen Grade System (B60/70). These mixtures were formulated in accordance with the Marshall mix design method and the Superpave mix design method. Rutting test also tested these mixtures to evaluate the performance of the bituminous mixtures under certain conditional parameters. The study clearly showed that the asphalt mixtures with binder PG70-10 are slightly performed better than the mixtures with binder (B60/70) In addition, the results obtained from Superpave mix design test have shown a better performance indicator than those obtained with the Marshall Stability tests and they were superior at all stages of tests. These results provide a guide line to improving the production of the hot mix asphalt (HMA) and a foundation for different paving mixes, by adopting PG binder and Superpave Gyratory compactor (SGC) instead of Marshall hammer and binder B60/70.

Keywords: Pavement · Performance · Climatic · Asphalt mix design
Local materials

1 Introduction

Asphalt mixture properties such as density and air voids are largely dependent on the type of the compaction method, selection of materials and method of mix design. These properties, in turn, have an influence over the pavement performance indicators such as rutting, fatigue, potholes and alligator cracks (Malunga et al. 2014). Therefore, this research aims to improve the local mix design formulation of the pavement asphalt mixture to obtain an economical mixture that would meet the requirements for the characteristics of the pavement in hot weather. The Marshall Mix Design Method selected the asphalt binder based on the empirical Penetration Grading Pen Grade System. It does not mandate for viscosity measurements and hence, it is difficult to establish the correct HMA. Moreover, temperature can be a major contributor to several types of distress (Salem et al. 2014). Binder and compaction method are perhaps the most important elements for an asphalt mixture design. They influence almost all the important properties of HMA mixtures such as stiffness, stability, durability, etc.

© Springer International Publishing AG, part of Springer Nature 2019
A. Sevi et al. (eds.), *Enhancements in Applied Geomechanics, Mining, and Excavation Simulation and Analysis,* Sustainable Civil Infrastructures,
https://doi.org/10.1007/978-3-319-95645-9_2

Therefore, the following research will adopt a new asphalt binder PG which should be related to the climatic conditions of Libya and a new method of compaction which is SGC. Marshall Hammer Compactor is very old method of compaction in the laboratory. Nevertheless, the variation in the methods of laboratory compaction is not only the result of the methods of assessment, but is due to the technology used (Foster 1982).

2 Objectives

This research aims to show the effect of some factors on the asphalt mix design such as the type of bitumen binder, skeleton of aggregate and the method of compaction. These mixtures were formulated in accordance with the Marshall mix design method and the Superpave mix design method. Rutting test also tested these mixtures to evaluate the performance of the bituminous mixtures under certain conditional parameters. Additionally, the purpose of this paper is to show the limitation of Marshall Mix Design Method, which is still used today and to increase the implementation of earlier researches.

3 Background and Literature Review

Hubbard-Field Method might be considered as the first formal design method for asphalt mixtures. It was originally developed to design sand-asphalt mixtures and later modified for aggregates (Roberte et al. 2002). The Marshall mix design method, initially developed by Bruce G. Marshall from the Mississippi Highway Department in 1939, and later amended by the US Waterways Experiment Station of the US Army to include deformation measurements. This is based on two criteria, a minimum stability and a minimum range of flow values. However, it was based on a single piece of equipment to design and control the asphalt paving mixtures which referee to as the Marshall Test. The Superpave mix design method, for Superior Performing Asphalt Pavement System, was developed in the 1980s as a part of the Strategic Highway Research Program (SHRP). Superpave addresses aggregate selection and asphalt binder selection in the perspective of future traffic and climatic requirements. It selects aggregates based on gradation distribution, angularity, clay content, water absorption, abrasion, soundness. The asphalt binder selection relies on the rutting analyser and the dynamic (AASHTO 1993). In Libya, mix design of asphalt mixture is based on the Marshall method which is an empirical design method that does not replicate the compaction, materials, and climate prevailing in Libya. Previous studies have shown that the compaction by impact in the Marshall method is unrealistic compared to the SGC, which simulates the field density. And the Pen Grade system is typically conducted at a single test temperature of 25 °C, which does not account for the entire temperature spectrum to which the asphalt-binders are subjected to in the field (Almadwi and Assaf 2017). However, the term "Superpave" refers to more than just the computer program. The system includes test equipment, test methods, and criteria. Most important, it represents an improved system for specifying component materials, asphalt mixture design and analysis, and pavement performance prediction (Asi and Khalayleh 2011).

4 Methodology

Two asphalt concrete mixtures of aggregate-asphalt binder were designed using two different binders, PG70-10 and B60/70. These mixtures were formulated in accordance with the Marshall mix design method and the Superpave mix design method. Rutting test also tested these mixtures to obtain an economical mixture that would meet the requirements for the characteristics of the pavement in hot weather. These specimens were used for a comparison under different Libyan weather conditions. Evidently, the test mixtures contained several asphalt bitumen contents both above and below the optimum content of asphalt.

4.1 Mix Design Experiments

The purpose of this research is to determine the proper proportions of aggregates and asphalt to obtain an economical mixture that would meet the requirements for the characteristics of the pavement in hot weather. Over the years, several design and development methods have been taking place and have been implemented by various agencies. This review focuses on the producing and design of HMA in Libya.

4.2 Aspects Influencing the Design of Asphalt Mixtures

The physical properties of the materials that are used in this mix design are shown in Table 2.

4.2.1 Aggregate Gradation

In general, stone skeleton in HMA, such as porous asphalt and SMA (Stone Matrix Asphalt), are quite resistant to rutting due to their high stone concentration. To obtain a good overall asphalt mixture performance, there must be an aggregate structure that favours resistance to rutting. The gradations of aggregates are expected to pass within specified bands presented on a semi-logarithmic (semi log) graph. Specifications, however do not address the more recent findings regarding the physical characteristics of aggregates. These specifications include findings from studies conducted by Kandhal et al. (1998), who found that Aggregate shape properties, such as form, angularity, and surface texture, highly influence the performance of HMA. These findings have led to the introduction of the Superpave consensus aggregate properties. They are aimed at improving performances of HMA mixes.

4.2.2 Type of Bituminous Binder

Asphalt binder changes its properties with time under traffic loads and environmental affects. Asphalt-binder B60/70 is used in the traditional Marshall Mix Design Method and is selected based on the empirical Pen Grade System. It does not mandate for viscosity measurements and hence, it is difficult to establish the correct HMA. Lastly, the original Pen Grade System was developed for low traffic loading conditions, which regrettably do not tally with the current high traffic on the roads in the hot climate of Libya. Due to these limitations, asphalt-binder selection using this method to ensure satisfactory pavement performance is highly questionable. In recent years, a new asphalt-binder selection criterion known as the Performance Grading system has been developed by the Strategic Highway Research Program (SHRP) to overcome of some the asphalt pavement distresses such as rutting, fatigue cracking, and thermal cracking The PG concept is based on the rheological theory that an asphalt-binder's property should be related to the climatic conditions under which it is to be used. Engineering properties believed to be related to the expected performance (maximum 7-day pavement temperature, 1-day minimum pavement temperature, loading duration based on truck speed, and traffic volume) are featured in the PG binder specifications to allow for a proper selection of asphalt roads (Wahhab et al. 1997).

4.2.3 Compaction Method

Compaction of asphalt mixes in flexible road surfaces play an important role in the behavior of the pavement under the traffic load. The properties of the mixture, such as density and air voids, strongly depend on the degree and the method of compaction. Marshall Hammer Compactor is the oldest method of compaction in the laboratory. The number of impacts applied to each face of the specimen was set at 35, 50 or 75, depending on the anticipated volume of traffic. The higher the amount of traffic, the greater the number of blows. In fact, a proper method of compaction is also necessary to take measures to extend the life of the pavement using various compaction methods. One of the compaction methods is the SGC. Although the equipment used in the Marshall method is inexpensive, studies have shown that the impact compaction is unrealistic compared to a SGC that simulates a field density of about 50% of the time (Button et al. 1994).

4.3 Materials Used in Hot Mix Design

4.3.1 Aggregate

In this research, the mix of coarse and fine aggregate with sizes between 0.08 to 28 mm was used in the design of asphalt mixture according to the requirements. The characterises and gradation for both coarse and fine aggregate were shown in Tables 1, 2, and Fig. 1.

Table 1. Sieve analysis result and aggregate gradation for laboratory mix design

Sieve size (mm)	% lower limit	% upper limit	% passing
28	100	100	100
20	95	100	98.7
14	67	90	85.6
10	52	75	68.1
5	34	55	44.87
2.5	24	45	27.6
1.25	16	39	17.7
0.63	9	31	12.7
0.31	6	23	9.68
0.16	4	15	7.55
0.08	3	8	6.33

Table 2. Specific gravity for the compound aggregates and asphalt binders

Properties of materials in mix design				
Agg. size and asphalt binder	% of Agg. in mixture	Bulk specific gravity	Compound mix specific gravity	% water absorption
20–14 (mm)	15	2.739	2.716	0.630
14–10 (mm)	15	2.736		0.670
10–5 (mm)	26	2.733		0.700
Crushed 0–5 (mm)	36	2.698		0.005
Sand 0–5 (mm)	3	2.595		0.007
Filler (mm)	5	2.700		0.006
Bitumen (B60/70)	4.50	1.020	1.020	–
Bitumen PG 70-10	4.25	1.290	1.029	–

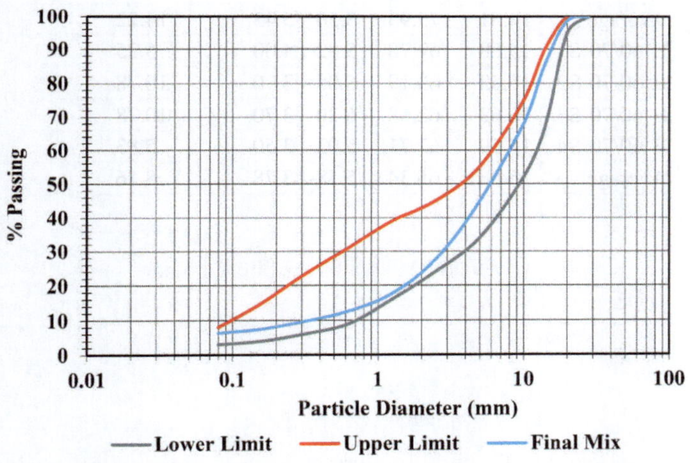

Fig. 1. Aggregate gradation specification

4.3.2 Bituminous Binder

In this research, the bituminous binder for asphaltic concrete mix was the bitumen of PG70-10 or B60/70.

4.3.3 Mineral Filler

The mineral filler used for this study was the dust of limestone. It should be dry enough and should be essentially free of agglomerations.

5 Laboratory Experiments Results

5.1 Marshall Mix Design Result

In this experiment, specimens were conducted by using bitumen binder B60/70 and the other specimens by using bitumen binder PG70-10. The specimens were then tested, and the results obtained are presented in Table 3, and illustrated in Fig. 2.

Table 3. Calculation of Marshall stability based on LC 26 Q.C standard

Calculation of Marshall stability					
Sample No.	VMA%	VFA%	Va%	Flow (mm)	Stability (KN)
PG70-10 S1	14.14	78.22	3.07	5.30	15.88
PG70-10 S2	14.84	73.90	3.87	6.40	14.36
PG70-10 S3	14.81	74.1	3.83	5.40	13.93
PG70-10 S4	14.25	77.49	3.20	4.60	12.74
Average	14.51	75.93	3.50	5.43	14.22
B 60/70 S1	15.58	67.78	5.02	4.00	5.25
B 60/70 S2	16.52	63.19	6.08	3.90	10.78
B 60/70 S3	16.63	62.68	6.20	3.70	10.78
B 60/70 S4	15.59	67.71	5.03	3.50	5.83
Average	16.08	65.34	5.58	3.78	8.16

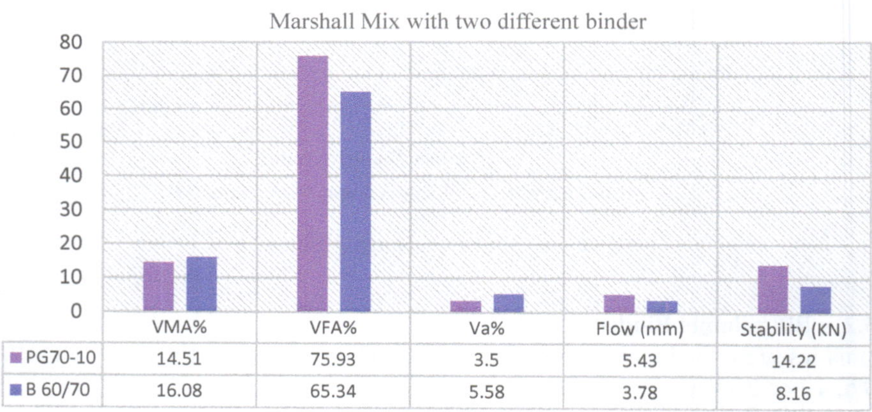

Fig. 2. Values of Marshall asphalt mixes with two different binders

5.2 Superpave Mix Design Test Result

In this experiment, specimens were tested by using bitumen binder B60/70 and the other specimens by bitumen binder PG70-10. The obtained results are shown in Table 4 and illustrated in Fig. 3.

Table 4. Mixtures properties from the (SGC), based on LC 26 Q.C standard

Result of SGC specimens with two different binders and Vbe = 12.28						
Numbers of gyrations	Binder PG (70-10)			Binder B (60/70)		
	VMA%	VFA%	Va%	VMA%	VFA%	Va%
10	22.01	49.60	11.10	24.34	43.53	13.75
80	18.15	63.15	6.69	17.62	65.50	6.10
200	15.05	79.02	3.16	15.00	79.50	3.08

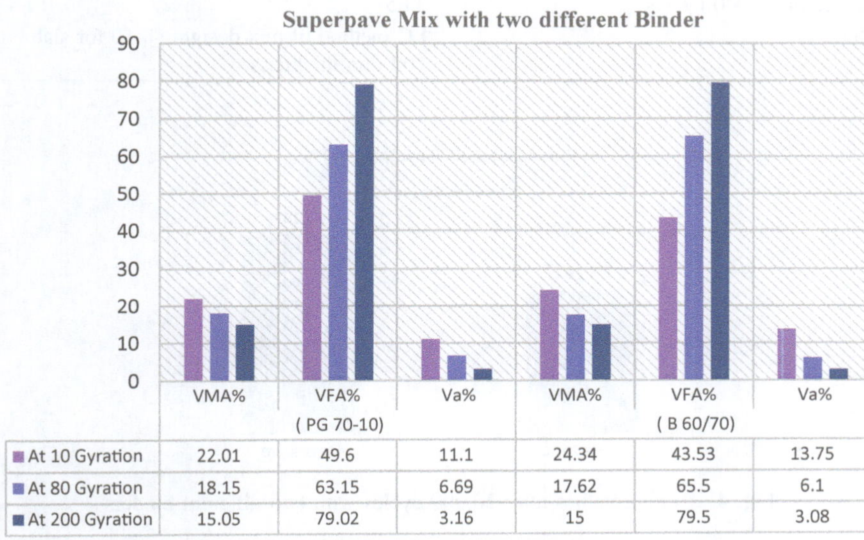

Fig. 3. Values of Superpave asphalt mixes with two different binders

5.3 Rutting Test Result

In this experiment, specimens had been tested by using bitumen binder B60/70 and the other specimens by bitumen binder PG70-10. The obtained results are presented in Table 5 and illustrated in Figs. 4 and 5.

Table 5. Results of rutting analyzer test

Number of cycles (cumulative) at 65 °C	Rutting (mm)			
	Specimens with PG70-10		Specimens with B(60/70)	
	S_I	S_{II}	S_I	S_{II}
1000	3.51	2.16	4.69	4.09
3000	3.98	2.60	5.26	5.02
10,000	4.12	2.90	5.90	5.86
30,000	4.42	3.28	7.48	7.07
Average of 30,000 cycles	3.85		7.27	
Criteria	LC method of mix design ≤ 10 for slab 100 mm thickness			

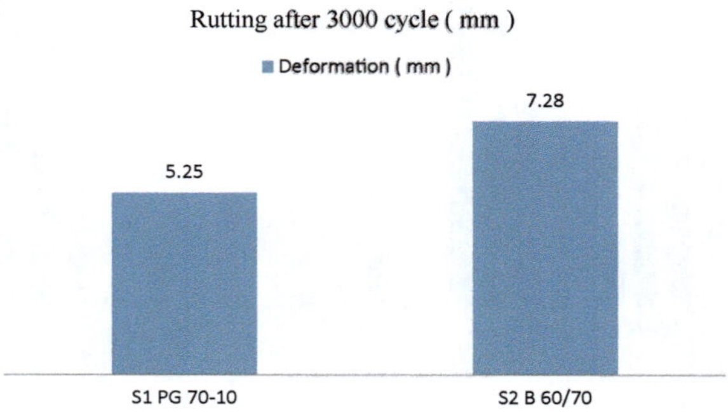

Fig. 4. Rutting values after 30,000 cycles with two different binders

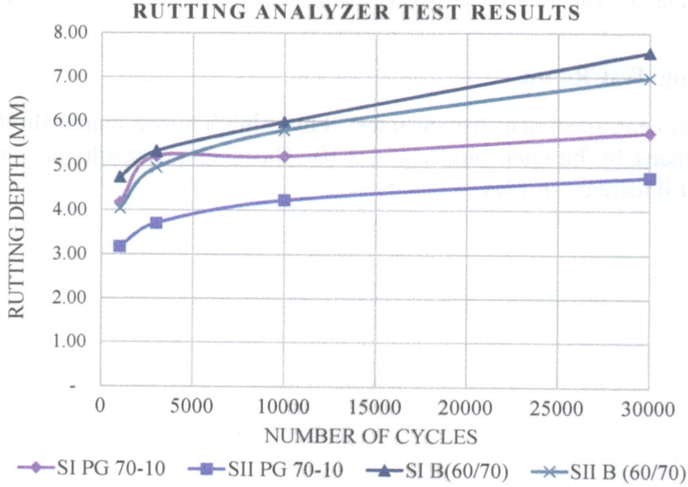

Fig. 5. Rutting depth of different specimens with two different binders

6 Synthesis and Discussion of the Findings

The asphalt binder that was used in the Marshall method may not be sufficient to ensure that selected asphalt binders can satisfactorily meet the paving service temperatures experienced in Libya. The method of selecting the binder and the bituminous content in the mixture is still questionable. However, in this study, the conventional local specifications of selecting and blending the aggregate and binder that were used in the Marshall mix design did not perform well and thus yielded inadequate results. On the other hand, the Marshall mixes were designed using the SGC, PG asphalt-binders and a proper aggregate gradation indicated that the Marshall method performed like the Superpave method. Asphalt-binder type, asphalt-binder content, aggregate properties, and gradation showed a significant effect on the performance of the mixes as expected. Tables 3, 4 and 5 illustrated all the laboratory experiment results. Figures 2, 3, 4, and 5, show the behavior of asphalt mixtures with new binder PG70-10 and a new compaction method under the hot weather condition such as in Libya. It can be seen from Figs. 2, 3, and Tables 3, 4 that the volumetric properties values of the Superpave mixes are properly better than that of Marshall mixtures. The flow values of Marshall mixtures with PG70-10 are slightly less than that of Marshall mixes with B60/70. This could be due to the differences in the binder properties and in the compaction techniques used. The SGC rotates at a constant rate during the compaction, and this characteristic provides around a better orientation of aggregate particles and aggregate interlock. This process simulates closely to the field compaction. In Figs. 4, 5 and Table 5, asphalt blends using the asphalt binder PG70-10 showed better resistance to rutting than the mixes using the asphalt binder B60/70. This study showed that in rutting test, most of the permanent deformation occurs in the upper four inches of HMA. The study clearly showed that the amount of air voids in the mixture was one of the most important properties. Although the number of air voids had a significant effect on the pavement deformation, this parameter could not be used to predict the deformation. When air voids were low, as a rule, there was considerable of deformation, and when air voids were high, it was much less likely of deformation. It is therefore important that air voids be carefully controlled during construction. In the future, the air voids will be determined from samples compacted with the SGC. Regardless of the type of compaction used, it is important that the sample is compacted to about the same density as in the field after several years of traffic.

7 Conclusions and Recommendations

The proper selection of the aggregates, method of compaction and the asphalt binder can improve the pavement performance. However, the most important factors which affects the pavement performance are the compaction technique, type of asphalt binder and selection of aggregate. In this research, two asphalt concrete mixes were designed using two different binders. These mixtures were formulated in accordance with the Marshall mix design method and the Superpave mix design method and these mixtures also tested with Rutting test. The analysis was conducted to evaluate HMA properties such as deformation, durability, air voids, voids in mineral aggregate, voids filled with asphalt, etc. The study in which the Marshall mixes were designed using the SGC, PG asphalt-binders and a proper aggregate gradation (source and consensus requirements), indicated that the Marshall method performed like the Superpave method. The type of asphalt-binder, the asphalt-binder content, and the aggregate properties, showed a significant effect on the performance of the mixes, as theoretically expected. A virtual need to evaluate the mix-design methods on the hot weather locale is still mandatory. It therefore can be concluded that the Marshall method can perform equally well if the method of compaction and asphalt-binder selection criteria are updated in Libya.

Acknowledgements. It is with great joy and lightness of spirit that I offer my deepest regards to my friends who have assisted and supported me in countless ways as I journeyed through the process of undertaking, creating, and completing this paper.

References

Almadwi, F.S., Assaf, G.J.: Performance testing of paving mixes for Libya's hot and arid conditions, using Marshall stability and SUPERPAVE gyratory compactor methods. In: International Congress and Exhibition "Sustainable Civil Infrastructures: Innovative Infrastructure Geotechnology", pp. 313–323. Springer, Cham (2017)

American Association of State Highway & Transportation Officials: AASHTO Guide for Design of Pavement Structures, 1993, vol. 1. AASHTO (1993)

Asi, I.M., Khalayleh, Y.: Adaptation of Superpave asphalt concrete mix design procedure to jordan climatic and traffic conditions. Int. J. Pavement Res. Technol. **4**(3), 154–161 (2011)

Button, J.W., Little, D.N., Jagadam, V., Pendleton, O.J.: Correlation of Selected Laboratory Compaction Methods with Field Compaction. Transportation Research Record, 1454 (1994)

Cominsky, R.J., Huber, G.A., Kennedy, T.W., Anderson, M.: The Superpave Mix Design Manual for New Construction and Overlays, No. SHRP-A-407. Strategic Highway Research Program, Washington, DC (1994)

Foster, C.R.: Development of Marshall Procedures for Designing Asphalt Paving Mixtures, No. IS 84. National Asphalt Pavement Association (1982)

Kandhal, P.S., Foo, K.Y., Mallick, R.B.: A Critical Review of VMA Requirements in Superpave, No. NCAT Report No. 98-1. National Center for Asphalt Technology (1998)

Kennedy, T.W., Huber, G.A., Harrigan, E.T., Cominsky, R.J., Hughes, C.S., Von Quintus, H., Moulthrop, J.S.: Superior Performing Asphalt Pavements (Superpave): The Product of the SHRP Asphalt Research Program, No. SHRP-A-410. Strategic Highway Research Program, National Research Council, Washington, DC, USA (1994)

Lanham, M.D.: Development of Marshall Procedures for Designing Asphalt Paving Mixtures, Information Series 84. Natl Asphalt Pavement Assoc. (1982)

Malunga, R., El Halim, A.A., Muya, M., Mwanza, A.D., Faruk, A.N., Mushota, C., Walubita, L. F.: Theoretical review of different asphalt mix-design methods and their applicability for developing countries like Zambia. In: Design, Analysis, and Asphalt Material Characterization for Road and Airfield Pavements, pp. 210–217 (2014). https://doi.org/10.1061/9780784478462.026

Roberte, F.L., Mohammad, L.N., Wang, L.B.: History of hot mix asphalt mixture design in United States. J. Mater. Civ. Eng. **14**(4), 279–293 (2002)

Salem, H.A., Uzelac, D., Matic, B.: Temperature zoning of Libya desert for asphalt mix design. In: Applied Mechanics and Materials, vol. 638, pp. 1414–1426. Trans Tech Publications (2014). https://doi.org/10.4028/www.scientific.net/AMM.638-640.1414
Wahhab, H.I.A.A., Asi, I.M., Al-Dubabe, I.A., Ali, M.F.: Development of performance-based bitumen specifications for the Gulf countries. Constr. Build. Mater. **11**(1), 15–22 (1997)

Review and Analysis on Using the Analytical Approaches for Predicting the Pavement Performance

Rajashree Tapase[1], Dilip Aldar[2], and Anand Tapase[2](✉)

[1] Annasaheb Dange College of Engineering and Technology, Ashta, Maharashtra, India
rajashree_tapase@rediffmail.com

[2] Rayat Shikshan Sanstha's Karmaveer Bhaurao Patil College of Engineering, Satara, Maharashtra, India
Dilip_aldar@rediffmail.com, tapaseanand@gmail.com

Abstract. The complex characteristics of the present day in pavement system are making the available design procedure impractical for recurring tasks. It is seen the pavement deteriorates not only due to the combined effects of traffic loading and environmental conditions but also its failure takes place due to deficiencies construction, materials, and maintenance. Predicting pavement performance before its actual execution is possible with the help of analytical tools once they are validated. The paper provides a state of the art review of different analytical approaches implemented for the analysis of pavements and evaluating its performance. From the available literature, it should be noted that the ANN and FEM approaches can be realistically applied which do not require a formulation or function of the solution. Such tool will accommodate not only the thickness design but will assist the decision makers in finding optimum strategies for providing, evaluating and maintaining pavements in a serviceable condition for the longer duration. In this connection, it should be noted that the application of artificial neural network (ANN) and finite element method (FEM) will help in predicting the performance of different design sections for new pavement construction as well as for the maintenance operations in the form of overlay design. Performance prediction prior to actual construction will help to set the maintenance budget at the network level by assigning most cost-effective strategy at the project level.

1 Introduction

The root causes of the poor performing pavements or earlier deterioration of roads in the form of potholes, undulations, fatigue cracking depends on the local factors and may differ to the pavement to pavement. Finding an empirical solution for its maintenance may result again in its premature failure or the solution may be highly not realistic for the particular condition. A number of researchers including Tapase and Ranadive (2016, 2017) have already reported that earlier deterioration of roads and its time-consuming rehabilitation/maintenance operation are the direct or indirect cause behind an increase in road fatalities at around 3% compared to a preceding year in

© Springer International Publishing AG, part of Springer Nature 2019
A. Sevi et al. (eds.), *Enhancements in Applied Geomechanics, Mining, and Excavation Simulation and Analysis,* Sustainable Civil Infrastructures,
https://doi.org/10.1007/978-3-319-95645-9_3

India. The World Health Organisation (WHO), Ministry of Road Transport and Highways (MoRTH) have already started taking actions to mitigate the fatalities and to strengthen road safety as well as improve ease of transport across the various parts within the country and worldwide. Yang et al. (2015) have observed number of overlayed sections of the Arkansas Interstate roads have experienced severe premature rutting and cracking and reported the need of pavement management tool for pavement maintenance actions.

It is difficult to encounter the existing problem of the deteriorated road in a well-timed and economical manner simultaneously, due to various complexities in the maintenance procedure. The properties of existing layers are needed to be evaluated as well as the complex interaction of different possible strategies needs to be forecasted for planning the maintenance, rehabilitation, and reconstruction of the deteriorated pavement. So, applying various trials at the network level with the help of analytical tool for predicting the performance of pavement before construction of new pavement or for maintenance or rehabilitation of deteriorated pavement will be helpful for selecting the best possible optimum strategies for providing, evaluating and maintaining pavements in a serviceable condition for the longer duration. In this connection, it should be noted that versatile finite element method and the artificial neural network holds a bright assurance. Present paper is a state of the art review focusing the advantageous application of various methods like finite element method (FEM) and the artificial neural network (ANN) in the civil infrastructures from failure to sustainability and sequentially a comprehensive sensitivity analysis method is proposed as reported in Fig. 1 to explore the influence of analytical analysis in design and planning the best possible optimum strategies for providing, evaluating and maintaining pavements in a serviceable condition for the longer duration.

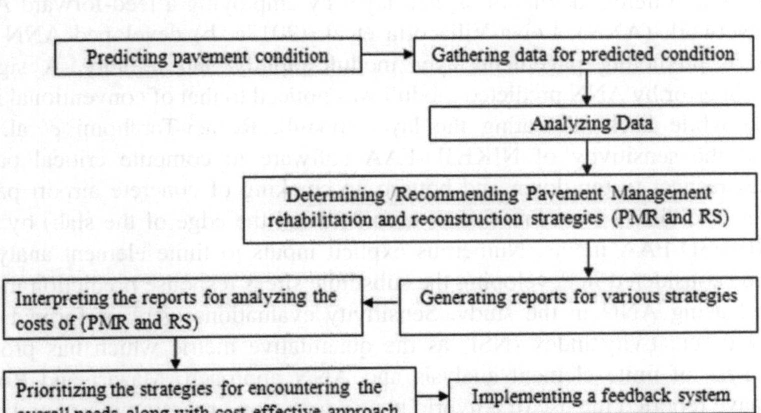

Fig. 1. General artificial neural network methodology

2 Literature Review

In recent years considerable analytical research efforts have been expended in civil engineering field due to the versatile methods like FEM, ANN and so on. The development of the mechanistic-empirical pavement design which is applied by a number of countries for the design of pavement is more realistic and has offered a potential for using more improved material characterization and constitutive models of both flexible and rigid pavements. A number of researchers have already documented the usefulness of FEM and ANN in predicting the performance of flexible as well as the rigid pavement. The Mechanistic Empirical Pavement Design Guide recommends JULEA a multilayer elastic program to compute flexible pavement responses (NCHRP 2007). The recently released IRC: 37-2012 guidelines for the design of flexible pavement recommends using the IITPAVE for layered system analysis.

Gogoi et al. (2013) studied to investigate if there is any co-relation in rutting and fatigue distress. Tapase and Ranadive (2016), reported the usefulness of two dimensional finite element analyses to study the effect of variation in thickness of different component layers on the critical parameters. Zheng et al. (2017) have developed genetic algorithm (GA)-back propagation (BP) neural network model based on the experimental data for evaluating the relationship between long-term skid resistance of epoxy asphalt mixture with the various engineering parameters for the running vehicles. Xiao and Amirkhanian (2009) as stated that the ANN model had higher accuracy then regression-based model for estimating the stiffness of rubberized asphalt concrete. Shafabakhsh et al. (2015) have reported the effectiveness of ANN method and its performance in evaluating the strengths of the hot mix asphalt mixtures. Ghanizadeh and Fakhri (2017) reported a nonlinear regression equation for determining equivalent frequencies at different depths of asphalt layer by employing a feed-forward Artificial Neural Network (ANN). Leiva-Villacorta et al. (2017a, b) developed ANN models capable of predicting pavement layer moduli rapidly and reliably. A significant decrease in error by ANN predicted moduli was noticed to that of conventional iterative approach while back-calculating the layer moduli. Rezaei-Tarahomi et al. (2017) evaluated the sensitivity of NIKE3D-FAA software to compute critical pavement responses respect to top-down and bottom-up cracking of concrete airport pavement structures at different locations (interior, corner, and the edge of the slab) by varying other NIKE3D-FAA inputs. Numerous explicit inputs to finite element analysis that need to be considered in developing the substitute stress response prediction model are evaluated using ANN in the study. Sensitivity evaluations were performed using a normalized sensitivity index (NSI) as the quantitative metric which has proven the effectiveness of finite element analysis and ANN approach. Mazari and Rodriguez (2016) have reported the use of a hybrid gene expression programming-neural network technique for predicting the international roughness index (IRI) which represents the pavement roughness. The focus of the reported study is to couple genetic programming and artificial neural network for IRI prediction on a dataset collected from the LTPP database. The developed algorithm showed reasonable performance for prediction of IRI using traffic parameters and structural properties of pavement. Yu et al. (2018) study reports the application of regression model and finite element modeling for evaluating more accurate modulus back-calculation by reducing noise information from in situ—measured asphalt pavement deflection basin.

3 Artificial Neural Network Methodology

The artificial neural network is a method which has an ability to solve the problem by information obtained from the past experience. The configuration of ANN is done by interconnecting the computational elements, through a learning process. There are three basic building blocks of the artificial neural network which are as network architecture, setting the weights and the activation function. The arrangement of neurons into layers and the pattern of connection within and in between layer are generally called as the architecture of the net. The neurons within a layer are found to be fully interconnected or not interconnected. The number of layers in the net can be defined to be the number of layers of weighted interconnected links between the particular slabs of neurons. If two layers of interconnected weights are present, then it is found to have hidden layers. There are various types of network architectures: feedforward, feedback, fully interconnected net, competitive net, etc. The method of settings the values for the weights enables the process of learning or training. The process of modifying the weights connections between network layers with the objective of achieving the expected output is called training a network. The activation function is used to calculate the output response of a neuron. The sum of the weighted input signals is applied with an activation to obtain the response. For neurons in the same layer, same activation functions are used. There may be linear as well as nonlinear activation functions. The nonlinear activation functions are used in a multilayer net.

To assist decision makers in finding optimum strategies for maintaining pavement in a serviceable condition nowadays pavement management system plays a vital role in providing a cost-effective approach. The future performance of pavement is evaluated in terms of the expected impact of maintenance and rehabilitation treatments that are proposed at network level activities of PMS (AASHTO guidelines, Huang). Network level and project level are the two main activities of PMS wherein the network level addresses the overall budget and planning services including maintenance and rehabilitation and the project level is where the specific decision on maintenance strategies and funding allocation are made. So it is learned that pavement performance if predicted at the planning stage itself or in other words at the network level then it would be convenient to propose the cost-effective maintenance budget at the network level which will help for assigning the most cost-effective strategies at the project level.

An analysis of the existing pavement to determine/predict the cause of deterioration is necessary to identify and to decide the rehabilitation or maintenance alternative available. It should be a full proof strategies which should adequately address the cause of deterioration whether it is normal or abnormal, whether the material used is durable, adequacy in drainage, the environmental factor, present and future traffic and the geometrical factor.

There is a large number of maintenance, rehabilitation, and construction alternative available for both flexible and rigid pavements, and recycling has increased that number of options. Surface seals such as aggregate seal and slurry seal combined with localized repairs are often used as preventive maintenance treatment for flexible pavement, and they are also used as rehabilitation treatments on lower volume flexible pavement when structural improvements are not required. Overlays have become more versatile by combining them with inter-layers such as fabric, milling full or partial widths, and recycling part of an existing layer prior to applying the overlay. Other types of rehabilitation and reconstruction for flexible pavements include the following:

1. Cold in-place recycling followed by a new surface;
2. Hot in-place recycling with or without an overlay;
3. Reworking and stabilization of foundation materials followed a new surface;
4. Partial or full depth removal and replacement;
5. Full depth recycling; and
6. Overlay with PCC.

Another type of rehabilitation for rigid pavement may include

1. A series of maintenance treatments such as partial depth patching, full depth repair, surface grinding and joint sealing often referred to as concrete pavement restoration (CPR)
2. Break and seat with a concrete/asphalt overlay and
3. Bonded or unbounded PCC overlay.

Pavement design is the project level activity where detailed engineering and economic considerations are given to alternate combinations of sub-base base, and surface materials which will provide adequate load carrying capacity. Factors that are considered include; material, traffic, climate, maintenance drainage, and life-cycle costs.

4 Observations and Future Research Direction

The number of researchers have reported the usefulness of analytical tool in pavement design and its analysis; more rigorous pavement design can be achieved by the application of analytical tool which makes the pavement engineer check variety of combination of material and thickness combinations for the particular location.

Pavement performance if predicted at the planning stage itself or in other words at the network level then it would be convenient to propose the cost-effective maintenance budget at the network level which will help for assigning the most cost-effective strategies at the project level. Such tool will accommodate not only the thickness design but will assist the decision makers in finding optimum strategies for providing, evaluating and maintaining pavements in a serviceable condition for the longer duration.

To evaluate existing structural conditions and to predict the pavement condition after 'n' number of years, a methodology is proposed with the help of artificial neural networks (ANNs), which will use the collected data from NDT tests like falling weight deflectometer (FWD) and will interpret accordingly. The structural properties of the existing pavements can be back-calculated using the artificial neural network. A neural networks methodology as mentioned in Fig. 1 will be used for back-calculation model proposed in this research. With an intention, it is proposed to develop specific models for conventional flexible and rigid pavements a state-of-the-art literature survey is carried out as reported in the present study. By varying the inputs as per the selected trials, the optimum combination can be obtained rigorously once the model is validated with existing models or with actual field conditions. Reduction in computation time may be obtained by the adoption of an ANN-based approach for back-calculating the structural properties.

Accuracy in research depends extensively on the quality of the field data collected. Hence, it is proposed to use the available collected data from the past case histories for validating and confirming its accuracy before its use. Also, the authenticity of the in situ data will be acquired before its use for getting more rigors results and research findings. The present work is proposed with a view to prepare guidelines or a set of procedures which will help the pavement engineers to plan the rehabilitation budget at network level itself.

References

Abo-Hashema, M.: Artificial neural network approach for overlay design of flexible pavements. Int. Arab J. Inf. Technol. **6**(2) (2009)

American Association of State Highway Officials AASHTO: AASHTO guide for the design of pavement structures. AASHTO, Washington, D.C. (1993)

Diefenderfer, B.K., Al-Qadi, I.L., Diefenderfer, S.D.: Model to predict pavement temperature profile: development and validation. J. Transp. Eng. ASCE **132**(2), 162–167 (2006)

Ghanizadeh, A.R., Fakhri, M.: Quasi-static analysis of flexible pavements based on predicted frequencies using Fast Fourier Transform and Artificial Neural Network. Int. J. Pavement Res. Technol. (2017). https://doi.org/10.1016/j.ijprt.2017.09.002

Gogoi, R., Das, A., Chakraborty, P.: Are fatigue and rutting distress modes related? Int. J. Pavement Res. Technol. **6**(4), 269–273 (2013)

IRC: 37-2012. Guidelines for the design of flexible pavements. Indian Roads Congress, New Delhi (2008)

Kamboozia, N., Ziari, H., Behbahani, H.: Artificial neural networks approach to predicting rut depth of asphalt concrete by using of visco-elastic parameters. Constr. Build. Mater. **158** (2018), 873–882 (2018)

Leiva-Villacorta, F., Vargas-Nordcbeck, A., Timm, D.H.: Non-destructive evaluation of sustainable pavement technologies using artificial neural networks. Int. J. Pavement Res. Technol. **10**(2), 139–147 (2017)

Masad, S., Little, D., Masad, E.: Analysis of flexible pavement response and performance using isotropic and anisotropic material properties. J. Transp. Eng. ASCE **132**(4), 342–349 (2006)

Mazari, M., Rodriguez, D.D.: Prediction of pavement roughness using a hybrid gene expression programming-neural network technique. J. Traffic Transp. Eng. **3**(5), 448–455 (2016)

Moroval, N., Serin, S., Terzi, S., Saltan, M.: Prediction of the pavement serviceability ratio of rigid highway pavements by artificial neural networks. J. Adv. Technol. Sci. **2**(1), 12–25 (2013)

NCHRP: Evaluation of mechanistic-empirical design procedure. National Cooperative Highway Research Program, NCHRP Project 1-37A, National Research Council, Washington, D.C. (2007)

Ranadive, M.S., Tapase, A.B.: Investigation of behavioral aspects of flexible pavement under various conditions by finite element method. In: Yang, Q., Zhang, J.-M., Zheng, H., Yao, Y. (eds.) Constitutive modeling of geomaterials. Springer, Berlin, pp. 765–770 (2013). https://doi.org/10.1007/978-3-642-32814-5_100

Ranadive, M.S., Tapase, A.B.: Parameter sensitive analysis of flexible pavement. Int. J. Pavement Res. Technol. (IJPRT). Elsevier, Special Issue on Sustainability on Pavement Engineering (2016a). https://doi.org/10.1016/j.ijprt.2016.12.001

Ranadive, M.S., Tapase, A.: Pavement performance evaluation for different combinations of temperature conditions and bituminous mixes. Innovative Infrastruct. Solutions **1**(40), 1–5 (2016b). https://doi.org/10.1007/s41062-016-0040-9. (Springer)

Rezaei-Tarahomi, A., Kaya, O., Ceylan, H., Gopalakrishnan, K., Kim, S., Brill, D.R.: Sensitivity quantification of airport concrete pavement stress responses associated with top-down and bottom-up cracking. Int. J. Pavement Res. Technol. **10**, 410–420 (2017)

Shafabakhsh, G.H., Ani, O.J., Talebsafa, M.: Artificial neural network modeling (ANN) for predicting the performance of nano-modified hot-mix asphalt mixtures containing steel slag aggregates. Constr. Build. Mater. **85**, 136–143 (2015)

Tapase, A., Ranadive, M.: Performance evaluation of flexible pavement using finite element method. ASCE GSP 266, Geo-China 2016: Material, Design, Construction, Maintenance, and testing of pavement, 9–17 (2016). https://doi.org/10.1061/9780784480090.002

Thube, D.T.: Artificial neural network (ANN) based pavement deterioration models for low volume roads in India. Int. J. Pavement Res. Technol. **5**(2), 115–120 (2012). ISSN 1997-1400

Xiao, F.P., Amirkhanian, S.N.: The artificial neural network approach to estimating stiffness behavior of rubberized asphalt concrete containing reclaimed asphalt pavement. J. Transp. Eng. **135**(8), 580–589 (2009)

Yang, H.H.: Pavement Analysis and Design, 2nd edn. Pearson Education Inc./Dorling Kindersley Publishing Inc. (2008)

Yang, S., Braham, A., Chowdhury, N., Hossain, Z.: Linking the field and lab performance of interstate pavements. In: Innovative Materials and Design for Sustainable Transportation Infrastructure—Selected Papers from the International Symposium on Systematic Approaches to Environmental Sustainability in Transportation, pp. 19–27 (2015). https://doi.org/10.1061/9780784479278

Yu, J., Xiong, C., Zhang, X., Li, W.: The durability of innovative construction materials and structures more accurate modulus back-calculation by reducing noise information from in situ–measured asphalt pavement deflection basin using regression model. Constr. Build. Mater. **158**(2018), 1026–1034 (2018)

Zheng, D., Quian, Z.D., Liu, Y., Liu, C.B.: Prediction and sensitivity analysis of long-term skid resistance of epoxy asphalt mixture based on GA-BP neural network. Constr. Build. Mater. **158**(2018), 614–623 (2017)

Grouting Material Development and Treatment of Water and Mud Inflow Caused by TBM Tunneling in Fault Zones

Jiwen Bai[1,2], Shucai Li[1], Rentai Liu[1(✉)], Xiao Feng[1], Peng Jiang[1], and Bingchuan Cheng[1]

[1] Geotechnical and Structural Engineering Research Center, Shandong University, Ji'nan 250061, Shandong, China
bjwxxyy@126.com, rentailiu@163.com

[2] Graduate School of Engineering, Nagasaki University, Nagasaki 8528521, Japan

Abstract. Water and mud inflow caused by TBM (Tunnel Boring Machine) tunneling in fault zones is a serious and troublesome geological obstacle in underground engineering. The common solving method is to disassemble the cutterhead and tunnel shield of TBM, which can result in the unstability of the surrounding rock in the process of disassembly. Grouting without disassembly is a practicable method to reinforce the surrounding rock and reduce water inflow. However, the common grouting materials are difficult to remain under the influence of fracture flow, and the grouting pressure makes the broken surrounding rock and tunnel shield much easier to consolidate together. In order to solve the above problems, a new type of grouting material GT-1 was developed. The GT-1 grouting material possesses the characteristics of high and early strength. The initial setting time of grouting slurry is 92–452 s, and the retention rate of slurry in the water flow velocity (0.4 m/s) is 95%. According to the analysis of GT-1 grouting material and the position of water and mud inflow, the arrangement of grouting holes, grouting parameters and safety control in the process of grouting are designed, the grouting parameters are adjusted based on the real-time deformation of surrounding rock. The grouting effect comprehensive evaluation system which includes physical exploration method, core-drilling method and excavation method is built to examine the grouting effect. The TBM excavated across fault zones smoothly, which possesses directing significance for similar projects.

Keywords: Water and mud inflow · TBM tunneling · GT-1 grouting material
Grouting design · Grouting effect comprehensive evaluation system

1 Introduction

Tunnel Boring Machine (TBM) is used as an alternative to drilling and blasting methods in hard rock, either shielded or open-type TBM can be used. The rock mass quality is generally favorable for tunneling, but the encounter of weak and water

© Springer International Publishing AG, part of Springer Nature 2019
A. Sevi et al. (eds.), *Enhancements in Applied Geomechanics, Mining, and Excavation Simulation and Analysis*, Sustainable Civil Infrastructures,
https://doi.org/10.1007/978-3-319-95645-9_4

bearing zones is normal, and sometimes leads to extreme challenges (Hassanpour et al. 2009, 2010; Zhang 1999).

In unstable fractured rock or water bearing zones, shielded hard rock TBM can be used, which erect concrete segments to support unstable tunnel walls behind the machine (Farrokh et al. 2012). However, the geological disasters often appear with the investigations and evaluations in the pre-construction phase. Many scholars have conducted a lot research on geological obstacle prevention and treatment, for example, Dammyr et al. (2017) summarized some of the extreme challenges encountered in Norwegian subsea road tunnels. Adverse rock mass behavior and/or sudden large water inflow at high pressure can be challenging to handle with open-face TBM, and found out that it can be hard to predict adverse rock mass behavior ahead of the face during tunneling. Liu et al. (2017) described the application of a comprehensive surface geophysical investigation of underground karst systems ahead of the tunnel face. Surface Electrical Resistivity Tomography (S-ERT), Transient Electromagnetic Method (TEM), Geological Drilling (Geo-D) and Three-dimensional Cross-hole Electrical Resistivity Tomography (3D cross-hole ERT) were applied to gain a comprehensive interpretation. To begin with, S-ERT and TEM are adopted to detect and delineate the underground. Khave (2014) toke Zagros Water Conveyance Tunnel in western Iran as the research object, which crossed a vast unconfined aquifer, a series of carefully controlled field experiments in a 100 m pilot study area along the tunnel were studied. Suitable field models are established that may be interpreted as being associated with either water or air-filled solution channels.

At present, most of studies are about the disaster-causing mechanism and geophysical investigation of underground in TBM tunneling. Whereas because of complexity and agnosticism of geology and the non-back characteristics of TBM, even shielded hard rock TBM can also be stuck by water and mud inflow in fault zones during tunneling. Considering the above occurred disasters, the rescue of TBM caused by water and mud inflow in Yinsong tunnel is taken as the research background. A new type of grouting material GT-1 is developed. The grouting holes, grouting parameters and safety control are designed. The grouting effect comprehensive evaluation system is built to examine the grouting effect.

2 Site Specification

The case presented in this paper is the Yinsong Tunnel, the main water supply line from Songhua River in Jilin Province in the northeastern China. The joint and fault region crossed by the line is mainly concentrated within the area of 1 km around stake mark K46+492–K47+492. The TBM was stuck when the TBM intersected the F23-2 fault in the position of stake mark K47+371. Figure 1 is the location and geology maps of the studied area. The F23-2 fault is mainly composed of mud, breccia, gravel and clay fraction, which possesses tremendous liquidity under the influence of underground water, and the water pressure ranges from 0.5 to 0.8 MPa. The total amount of water and mud inflow is 120 m^3. However, the surrounding rock around the cutterhead and tunnel shield of TBM is still unstable. The common solving method is to disassemble the cutterhead and tunnel shield of TBM, which has a great risk about the unstability of surrounding rock in the process of disassembly (Barton 2000). Grouting without disassembly is a practicable method to reinforce the surrounding rock and plug water. However, the common grouting materials are difficult to remain under the influence of fracture flow, and the grouting pressure makes the broken surrounding rock and tunnel shield much easier to consolidate together (Barla and Pelizza 2000; Maidl et al. 2008).

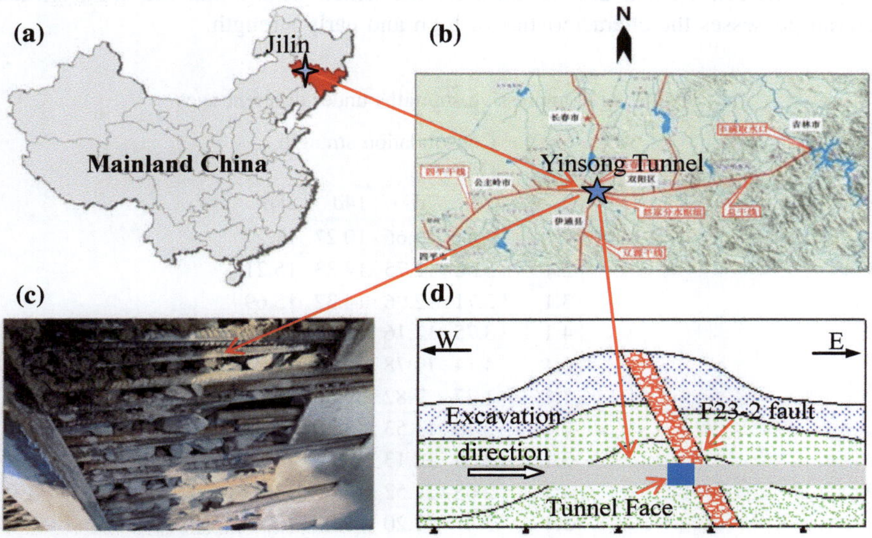

Fig. 1. Location and geology maps of the studied area. **a** Location of Jilin province in China. **b** General arrangement diagram of the water supply line from Songhua River, the five-pointed star stands for the location of the studied area—Yinsong Tunnel. **c** The picture of water and mud above the shield. **d** The relative position of tunnel face and F23-2 fault

3 Development of GT-1 Grouting Material

The most commonly used grouting material is neat cement, but it is difficult to remain under the condition of flowing underground water (Liu et al. 2011). Based on the requirements of stuck TBM project site and grouting, a new type of grouting material (GT-1) was developed. The GT-1 grouting material consists of cement and admixture (90% of sodium silicate, 5% of polycarboxylic acid water-reducer and 5% of polyethylene glycol), and the main performance parameters and characteristics under dynamic water condition were tested.

3.1 Test of Consolidation Strength

The consolidation strength of grouting material determines the sealing capacity of water and mud inflow and the long term stability of surrounding rock. In order to test the strength of GT-1 grouting material, the compressive strength under different ages were tested according to the testing standard GBT 1346-2011. For each water and mixture (cement and GT-1) ratio (W:C), the different volume ratio of cement (Vc) and admixture (Va) were tested separately.

As it is shown in Table 1, the compressive strength of GT-1 grouting material decreases with the increase of water cement rate, and increases with the increase of volume ratio of cement and admixture. In addition, the distinction of compressive strength between 7d and 28d is not obvious, which means that the GT-1 grouting material possesses the characteristics of high and early strength.

Table 1. Compressive strengths under different ages

W:C	Vc:Va	Consolidation strength (MPa)			
		1d	7d	14d	28d
0.8:1	1:1	2.54	9.66	10.27	11.27
	2:1	3.58	12.75	14.33	15.21
	3:1	3.71	12.96	14.32	15.69
	4:1	3.75	12.16	13.81	14.51
	5:1	4.14	11.78	13.14	14.03
1:1	1:1	1.77	7.82	9.06	9.74
	2:1	2.19	11.53	13.38	14.03
	3:1	2.34	12.13	13.84	14.52
	4:1	2.68	11.52	12.69	13.29
	5:1	3.47	11.20	12.09	12.96
1.2:1	1:1	1.13	6.19	8.69	9.27
	2:1	1.56	10.24	12.57	13.51
	3:1	1.87	10.19	11.88	13.48
	4:1	1.91	9.04	9.94	11.02
	5:1	1.89	7.73	8.68	9.79

3.2 Test of Setting Time

The initial/final setting time of grouting material determine the workability of slurry. Combined with the engineering field, the test is conducted under normal room temperature (20 °C). The initial/final setting time of grouting material were tested in the condition of different water mixture (cement and GT-1) ratio and volume ratio of cement (Vc) and admixture (Va).

The setting time of GT-1 grouting material are shown in Fig. 2. The initial setting time ranges from 92 to 452 s, and the final setting time ranges from 26 to 182 min. The initial/final setting time increase with the decrease of water cement ratio, and the initial/final setting time decrease with the increase of volume rate of cement and admixture. The setting time can be changed in a large time range, which can ensure the anti-dispersion of GT-1 grouting material in the condition of flowing underground water.

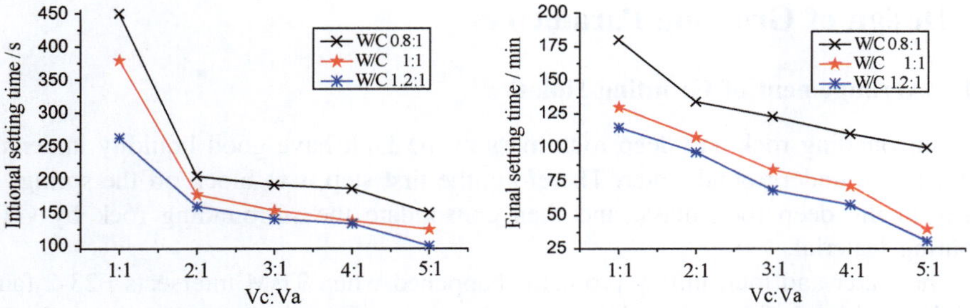

(a) Initial setting time (b) Final setting time

Fig. 2. Curves of initial/final setting time with W/C and Vc/Va

3.3 Test of Slurry Retention Rate

The slurry retention rate was tested at a water mixture (cement and GT-1) ratio (1:1) and varying volume ratio of cement and admixture (3:1 and 5:1). A certain amount of slurry after initial setting time are separately placed in the water channel, the mass (m1) is measured in the condition of hydrostatic condition, and mass (m2) is measured in the condition of flowing water. The slurry retention rate is defined as: P = m2/m1.

As it is shown in Fig. 3, the GT-1 grouting material processes high retention rate in the condition of flowing water. When the flow velocity of underground water is less than 0.4 m/s, the slurry retention rate remains more than 95%, and when the flow velocity of underground water is more than 0.4 m/s, the loss of slurry is relatively obvious. Hence the ideal environment of GT-1 grouting material should be water bearing fracture zones with inflows of less than 0.4 m/s.

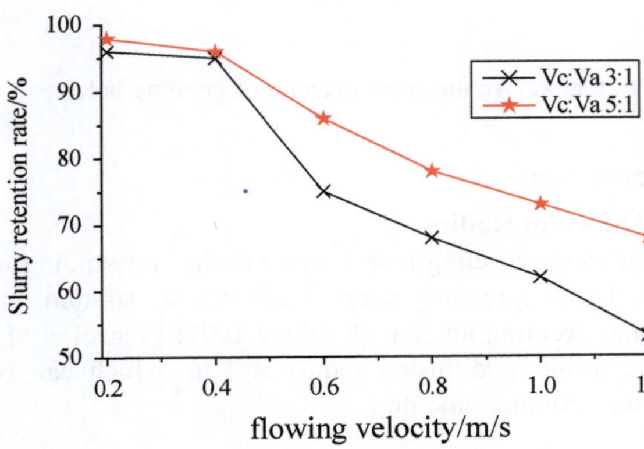

Fig. 3. Variation of slurry retention rate with water velocity

4 Design of Grouting Parameters

4.1 Arrangement of Grouting Holes

The surrounding rock and deep rock mass in the fault have good liquidity under the influence of underground water. Therefore, the first step is to block off the springs of water in the deep rock mass, and then consolidate the surrounding rock by GT-1 grouting material.

The water and mud inflow problems happened when TBM intersects F23-2 fault, and the position is focused on the upper left position. Therefore, the grouting holes are designed in Fig. 4. There are 6 groups of grouting holes, the upper 2 groups of grouting holes are designed to block off the springs of water in the deep rock mass, and the surplus groups of grouting holes are designed to consolidate the surrounding rock. In the first phase the GT-1 grouting material can block off the underground water effectively by controlling the initial setting time, and in the second phase GT-1 slurry spread range can be controlled by regulation of grouting pressure and setting time, which can protect the cutterhead and shield of TBM.

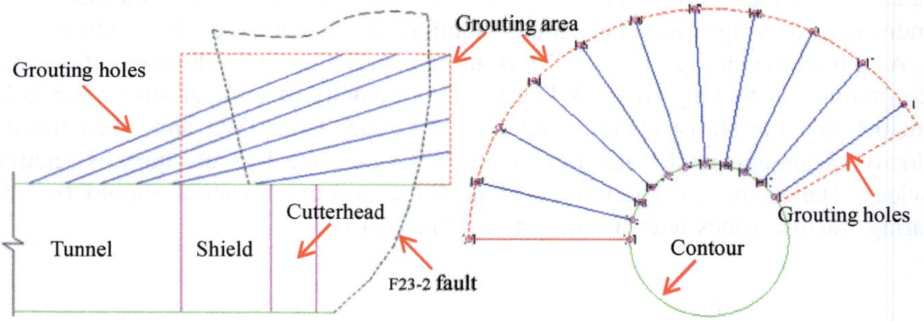

Fig. 4. Arrangement diagram of grouting holes

4.2 Grouting Parameters

4.2.1 Grouting Diffusion Radius

The grouting area of every grouting hole is close to the surrounding rock in the radial direction (3–10 m). Grout spreading range is affected by stratum, grouting material, grouting pressure and grouting amount (Robbins 1997; Frenzel et al. 2008). For this study, the designed grouting diffusion radius is 3 m, which can be controlled by grouting pressure and grouting amount.

4.2.2 Ending Grouting Pressure

The ending grouting pressure is 1–2 times of gravity of overlying strata for shallow tunnels (Zhang and Fu 2007). Therefore, the ending grouting pressure is 2–4 MPa. Taking the special geological conditions of fault and the shield safety of TBM into consideration, the ending grouting pressure should be flexible in the range of 2–4 MPa.

4.3 Deformation Control of Surrounding Rock

The deformation of surrounding rock directly reflects the stability of surrounding rock and compression load deflection of shield of TBM in the process of grouting. Therefore, the deformation monitoring of surrounding rock is very important, and the grouting parameters can be readjusted according to the deformation of surrounding rock. The deformations of roof strata 0–2 and 0–5 m are monitored.

As can be inferred from Fig. 5, The maximum value of relative deformation of roof in the range of 0–2, 0–5 m are 5.1, 8.2 mm, respectively. In the first phase the grouting area is the deep rock mass. Therefore, the relative deformations of roof in the range of 0–2, 0–5 m are not affected. In the second phase the relative deformation of roof in the range of 0–2, 0–5 m increase, and the relative deformation in the range of 0–5 m is more affected than the relative deformation in the range of 0–2 m. The relative deformation of roof is greater than 2 mm within one single grouting hole, grouting should be stopped. The surrounding rock is stable in the process of grouting, which ensure the safety of TBM.

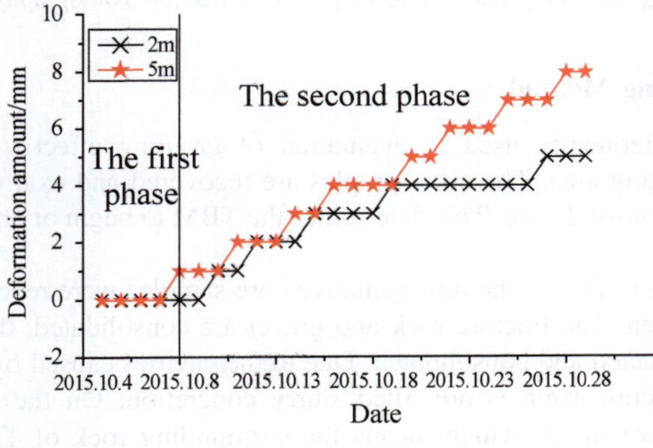

Fig. 5. The diagram of deformation of roof

5 Grouting Effect Comprehensive Evaluation System

5.1 Physical Exploration Method

True Reflection Tomography (TRT) was used to image changes in rock mass conditions ahead of the tunnel excavation up to 100 m. The technique of seismic pre-stack depth migration is used to correctly implement the complex hydrologic and structural imaging.

As it is shown in Fig. 6a, the grouting area is F23-2 fault which is in the position of stake mark K47+371. It can be appreciated from Fig. 6b, the s-wave velocity after grouting is larger than the wave velocity before grouting, which means the integral degree of rock mass is improved. Figure 6c shows that the grouting area 1 is water-rich area before grouting, and the water-rich area is continuous, by grouting treatment the negative reflection area 1 becomes discontinuous, which means the fault zones are reinforced and compacted sufficiently.

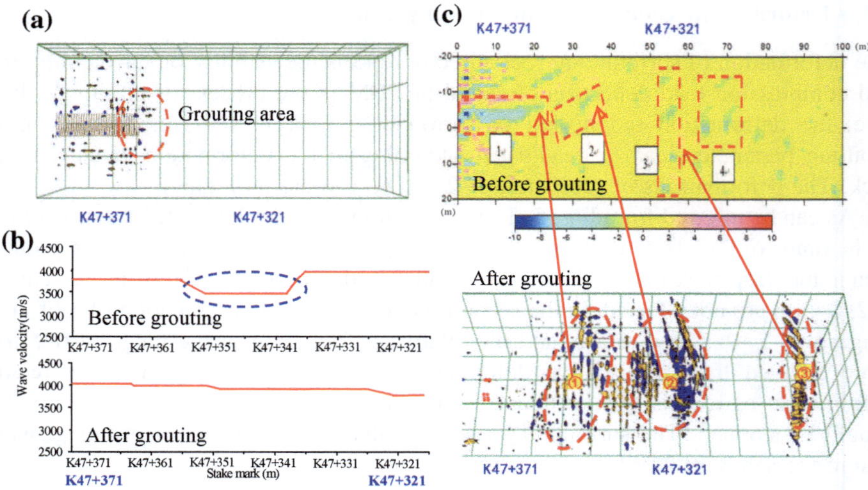

Fig. 6. The grouting effect by True Reflection Tomography

5.2 Core-Drilling Method

Core-drilling is frequently used in evaluation of grouting effect where the coring includes the grouting area. The core samples are recovered and examined for mineral percentages and consolidation. This determines the TBM to begin or abandon tunneling in grouting area.

As it is shown in Fig. 7, the representative core samples were recovered uniformly in the grouting area. The fracture rock and gravel are consolidated, the mud and clay fraction are compacted and consolidated. Due to the cavities caused by water and mud inflow, some of core samples are filled slurry concretion. On the whole, the core samples are relative intact, which means the surrounding rock of TBM tunnel after grouting is stable.

Fig. 7. The core samples recovered in the grouting area

5.3 Excavation Method

As it is shown in Fig. 8, the surrounding rock in the roof and face around cutterhead are consolidated well, there are no water penetration and slabbing of rock. The surrounding rock possesses better self-stability, which meets the requirements of TBM tunneling.

Fig. 8. The surrounding rock after grouting near the cutterhead

6 Conclusions

Generally speaking, uncertainties always prevail in predicting the geological conditions. Especially for the TBM tunneling, because of the non-back characteristics of TBM, even shielded hard rock TBM can be stuck by water and mud inflow in fault zones. In this research a series of tests were conducted to rescue the TBM by grouting instead of disassembling TBM. The following conclusions were determined from the findings of this research:

(1) A new type of grouting material GT-1 was developed. The grouting material possesses the characteristics of high and early strength. The initial/final setting time can be readjusted according to the water flow velocity, and the retention rate of slurry in the water flow velocity (0.4 m/s) is 95%.
(2) The arrangement of grouting holes, grouting parameters and safety control in the process of grouting are designed. The grouting holes are designed to block off the springs of water in deep rock mass and consolidate the surrounding rock, and the grouting parameters are adjusted based on the real-time water inflow and deformation of surrounding rock.
(3) The grouting effect comprehensive evaluation system which includes physical exploration method, core-drilling method and excavation method is built to examine the grouting effect. The surrounding rock processes better self-stability, which meets the requirements of TBM tunneling.

Acknowledgements. The authors would like to acknowledge the financial support from the National Basic Research Program of China (973 Program, Grand No. 2013CB036000), the National Natural Science Foundation of China (Grant No. 51479106).

References

Barla, G., Pelizza, S.: TBM tunnelling in difficult ground conditions. In: ISRM International Symposium. International Society for Rock Mechanics (2000)

Barton, N.R.: TBM Tunnelling in Jointed and Faulted Rock. CRC Press (2000)

Dammyr, Ø., Nilsen, B., Gollegger, J.: Feasibility of tunnel boring through weakness zones in deep Norwegian subsea tunnels. Tunn. Undergr. Space Technol. **69**, 133–146 (2017)

Farrokh, E., Rostami, J., Laughton, C.: Study of various models for estimation of penetration rate of hard rock TBMs. Tunn. Undergr. Space Technol. **30**, 110–123 (2012)

Frenzel, C., Käsling, H., Thuro, K.: Factors influencing disc cutter wear. Geomech Tunn. **1**(1), 55–60 (2008)

Hassanpour, J., Rostami, J., Khamehchiyan, M., et al.: Developing new equations for TBM performance prediction in carbonate-argillaceous rocks: a case history of Nowsood water conveyance tunnel. Geomech. Geoeng. Int. J. **4**(4), 287–297 (2009)

Hassanpour, J., Rostami, J., Khamehchiyan, M., et al.: TBM performance analysis in pyroclastic rocks: a case history of Karaj water conveyance tunnel. Rock Mech. Rock Eng. **43**(4), 427–445 (2010)

Khave, G.J.: Delineating subterranean water conduits using hydraulic testing and machine performance parameters in TBM tunnel post-grouting. Int. J. Rock Mech. Min. Sci. **70**, 308–317 (2014)

Liu, R., Li, S., Zhang, Q., et al.: Experiment and application research on a new type of dynamic water grouting material. Chin. J. Rock Mech. Eng. **30**(7), 1454–1459 (2011)

Liu, B., Liu, Z.Y., Li, S.C., et al.: Comprehensive surface geophysical investigation of karst caves ahead of the tunnel face: a case study in the Xiaoheyan section of the Water Supply Project from Songhua River, Jilin, China. J. Appl. Geophys. **144**, 36–49 (2017)

Maidl, B., Schmid, L., Ritz, W., et al.: Hardrock Tunnel Boring Machines. Wiley (2008)

Ribacchi, R., Fazio, A.L.: Influence of rock mass parameters on the performance of a TBM in a gneissic formation (Varzo Tunnel). Rock Mech. Rock Eng. **38**(2), 105–127 (2005)

Robbins, R.J.: Hard rock tunneling machines for squeezing rock conditions. In: Golser, Hinkel, Schubert (eds.) Three Machine Concepts. Tunnels for People. 23rd Assembly of the ITA, Vienna, pp. 633–638

Zhang, J.J.: The application and some problems of TBM and its prospects. Chin. J. Rock Mech. Eng. **18**(3), 363–367 (1999)

Zhang, J., Fu, B.: Advances in tunnel boring machine application in China. Chin. J. Rock Mechan. Eng. **26**(2), 226–238 (2007)

Traffic Data Characterization for Road Rehabilitation: A Case Study of the Korogwe-Mombo Road Section in Tanzania

Julius J. Komba[1(✉)], Mussa Mataka[2], John T. Malisa[2], Lubinda F. Walubita[3], and James W. Maina[4]

[1] Council for Scientific and Industrial Research (CSIR), University of Pretoria, P.O. Box 395, Pretoria 0001, South Africa
jkomba@csir.co.za

[2] Tanzania National Roads Agency (TANROADS), Central Materials Laboratory, P.O. Box 9452, Dar es Salaam, Tanzania
mussamataka@yahoo.co.uk, malisajt@yahoo.com

[3] TTI—The Texas A&M University System, College Station, TX, USA
lfwalubita@tamu.edu

[4] University of Pretoria, Lynnwood Road, Hatfield 0002, South Africa
james.maina@up.ac.za

Abstract. Traffic loading is one of the key inputs for the structural design of pavements. For pavement design purposes, heavy vehicles are mostly used for the estimation of the traffic load spectra, as they cause the most structural damage to pavements. As part of the pavement design process, heavy vehicle volume and axle load surveys are typically carried out to assist with the accurate estimation of the cumulative traffic loading over a pavement design period. However, traffic volumes, axle loads and, ultimately, the cumulative traffic loading are often not uniform due to factors, such as varying motorist population and economic activities along the length of road. In this paper, a comparative assessment of traffic loading estimated during the rehabilitation design in 2005 and the actual measured site-specific traffic loading in 2015, as well as the projected future traffic loading were conducted for the Korogwe-Mombo road section along the T2 trunk road that connects Tanzania's business hub of Dar es Salaam with the northern regional cities. The T2 trunk road is also used by heavy vehicles travelling to and from the neighbouring countries of Kenya and Uganda. This study found inter alia that the cumulative traffic loading based on the 2015 measured site-specific traffic data is approximately 2.8 times higher than the design traffic loading based on the 2005 traffic data, which illustrates the importance of using the latest, most accurate and reliable traffic data during rehabilitation design. To improve the accurate determination of traffic loading, traffic studies should ideally be conducted over a long period (typically over one year), but this is not practical and cost effective when traditional manual methods are used. Hence, road agencies should consider installing portable or permanent automated traffic and Weigh-In-Motion (WIM) monitoring systems.

Keywords: Pavement · Traffic · Weight · Axle load · Weigh-In-Motion (WIM)

© Springer International Publishing AG, part of Springer Nature 2019
A. Sevi et al. (eds.), *Enhancements in Applied Geomechanics, Mining, and Excavation Simulation and Analysis,* Sustainable Civil Infrastructures,
https://doi.org/10.1007/978-3-319-95645-9_5

1 Introduction

The primary objectives of pavement structure are to provide smooth riding quality to vehicles and protect the weaker (lower) layers such as subgrade against the effect of traffic loading. Although pavement deterioration may be caused by other factors such as climatic conditions, heavy vehicle volumes and axle loads are considered to be the primary causes of pavement deterioration. As such, heavy vehicle traffic loading has traditionally been used as the key input for the structural design of pavements (TRH 4 1996; MOW 1999). Pavement damages caused by heavy vehicles depend not only on the Gross Vehicle Mass (GVM), but also on the mass (weight) distribution onto the pavement. The later in turn depends on several factors, such as the number of axles on the vehicle, axle and wheel configuration, as well as axle load, tyre inflation pressure and contact stress (De Beer et al. 1997; Al-Qadi and Wang 2009; Greene et al. 2009; THM 14 2014). As part of pavement design process, heavy vehicle volume and axle load surveys are conducted to assist with the estimation of the future traffic loading over a pavement design period—typically 20 years for most flexible pavements (TRH 16 1991; TANROADS 2003; TRL 2004). The estimated traffic loading is used to determine the appropriate pavement design by taking into account the available material type and climatic conditions. However, traffic volumes, axle loads, and ultimately, the cumulative traffic loading are often not uniform due to varying motorist population and economic activities along the length of the road.

In 2005, the Tanzania National Roads Agency (TANROADS) identified a need for the rehabilitation of the Korogwe-Mombo road section (40 km long). The Korogwe-Mombo road section forms part of the North-East corridor (T2 trunk road), which is the main trunk road that connects Tanzania's East Coast (including Dar es Salaam city and Tanga Port) with the northern regional cities of Kilimanjaro and Arusha. The T2 trunk road is also the main route that links Dar es Salaam and Nairobi, the major trade centres of Tanzania and Kenya, respectively. The rehabilitation design of the road section was undertaken in 2006, based on traffic counts and axle load surveys carried out in 2005 (TANROADS 2006). The rehabilitation construction was initially planned to be completed in 2008. However, financial constraints delayed the construction works that only started in 2012 and were completed in 2014. Follow-up traffic counts and axle load surveys was carried out in 2015 after the road had been opened to traffic for a period of approximately one year. Together, the 2005 and 2015 traffic counts and axle load surveys constitute the basic data for the assessments conducted in this paper.

The objective of the current study was to conduct a comparative assessment of the initial traffic loading estimated during the rehabilitation design phase in 2005, the actual traffic loading determined after construction had been completed and the road had been opened to traffic in 2015, as well as the projected future traffic loading. The assessment provided an indication of the adequacy of the designed pavement and its expected future performance, and also highlighted the importance of measuring/collecting site-specific and accurate traffic volume and loading data.

2 Overview of Traffic Counts and Axle Load Surveys

2.1 Traffic Counts

Traffic counts for pavement design purposes can be performed by manual counts or automatic count methods such as the use of pneumatic tube counters. Each traffic count method requires different levels of effort and cost, and yields different levels of traffic detail. Regardless of the traffic count method used, the main purpose of traffic counts for pavement design is to obtain estimates of the base year traffic volumes. The actual composition of vehicles on a specific road section varies significantly, ranging from light passenger vehicles to buses and heavy vehicles transporting commercial goods. For pavement design purposes, the heavy vehicles portion of the traffic stream is commonly used, because light vehicles are considered to cause insignificant damage to the pavement (MOW 1999; TRH 4 1996; TRH 16 1991). During traffic counts, heavy vehicles are grouped into different categories to facilitate the determinations of traffic loading and their contribution to pavement damage (as will be demonstrated later in this paper). For example, the current practice in Tanzania requires that heavy vehicles be grouped into four different categories, namely Medium Goods Vehicles (MGV), Heavy Goods Vehicles (HGV), Very Heavy Goods Vehicles (VHGV), and Buses (see Table 1) (TANROADS 2003).

Table 1. Definition of heavy vehicle categories (TANROADS 2003)

Heavy vehicle categories	Definition
Medium goods vehicles (MGVs)	2 axles, including steering axle, and 3 tons empty weight or more
Heavy goods vehicles (HGVs)	3 axles, including steering axle, and 3 tons empty weight or more
Very heavy goods vehicles (VHGVs)	4 or more axles, including steering axle, and 3 tons empty weight or more
Buses	Seating capacity of 40 or more

Manual traffic counts are carried out by observers at a carefully selected observation point or counting station along a road section. The traffic count survey is usually a classified count whereby each vehicle passing an observation point is recorded on a prepared sheet/form according to the vehicle type, and each travel direction is recorded separately. The manual traffic count is usually undertaken over a short period (typically seven days), as it is not practical and cost effective to undertake manual traffic counts over a long period (i.e. 24-hour throughout the year). To improve the reliability of the traffic data counts, it is recommended that shorter period traffic counts be undertaken during normal days (i.e. days on which traffic patterns are not significantly affected by public and school holidays, or other events) (TMH 14 2014; TMH 3 2015).

Automatic traffic count an alternative to the manual traffic count can measure traffic volumes continuously over a long period and allow for capturing affects such as the seasonal variation of traffic volumes. The commonly used automatic traffic count systems can be grouped into three broad types namely: pneumatic tube, magnetic wire loops and piezo systems (TRL 2004). It is important to recognize that each of the available automatic traffic counting technologies has certain limitations that must be taken into account when establishing a traffic count station. Magnetic wire loop systems are the commonly used automatic traffic count systems in Southern Africa and are generally classified into two categories namely: intrusive loops (embedded or placed on the road pavement typically for long period traffic counts) and non-intrusive loops (placed on pavement surface and suitable for shorter period traffic counts) (TMH 3 2015). Figure 1 show photos of typical intrusive and non-inductive traffic count installations.

(a) Intrusive loops (b) Non-intrusive ("stick on") loops

Fig. 1. Typical intrusive and non-intrusive loop systems

2.2 Axle Load Surveys

Traditionally, static weighing that use a fixed weighbridge or a portable weighpad (see Fig. 2a) has been commonly used for the measurement of heavy vehicle axle loads. Alternatively, the measurement of heavy vehicle axle loads can be performed by Weigh-In-Motion (WIM) systems (FHWA 2013; TMH 14 2014). The bending plates (see Fig. 2b) are some of the most widely used in WIM systems. It should, however, be mentioned that axle load measured using WIM systems involves of static and dynamic load components (TMH 14 2014). Most of the available pavement design methods make use of static axle load data, and as such WIM axle load data may need to be processed to eliminate the dynamic load component.

(a) Portable static weigh pads (b) WIM bending plates

Fig. 2. Typical portable static weigh pads and WIM bending plates

Heavy vehicle axle loads are also affected by other factors such as changes to the legal axle load limits, the level of enforcements, as well as mechanical design and load carrying capacity of the vehicle. Table 2 shows the legal axle load limits for Tanzania (Government Notice No. 30 2001), alongside the legal axle load limits for South Africa (Government Notice R225 2000) and the East African Community (EAC) Vehicle Load Control Act (2013). It can be seen that in certain axle type/groups, the maximum permissible axle load limits for Tanzania are slightly higher than those for South Africa and the East African Community Vehicle Load Control Act. It is further noted that the maximum permissible load limit for the traditional dual tyres triple axle non-steering configuration (12 tyres) is the same as that of triple axle wide-base single (super single) tyres [6 tyres] for Tanzania (i.e., 24 tons), which may cause accelerated pavement damage. On the other hand, the East African Community Vehicle Load Control Act has reduced the legal load to 22.5 tons for triple axle wide-base single tyres to compensate for the damaging effects of the wide-base tyres, while South Africa does not encourage the use of wide-base single tyres (i.e. use 24 tons legal load for triple axle non-steering [12 tyres], regardless of whether they are normal or wide-base tyres).

Table 2. Legal axle load limits for Tanzania, South Africa and East African Community (EAC)

Type of axle/axle group	No. of tyres	Maximum permissible load on axle/axle group (tons)		
		Tanzania	South Africa	EAC
Single steering drive operated	2	8	7.7	–
Two steering drive operated	4	14	–	–
Single steering draw bar controlled	4	9	–	–
Single non-steering	2	8, 10, 12	8	8
Single non-steering	4	10	9	10
Tandem non-steering	4	12	16	–
Tandem non-steering	6	15		–
Tandem non-steering	8	18	18	18
Tandem steering (dolly)	8	16	–	–
Triple non-steering	10	21	–	–
Triple non-steering	12	24	24	24
Triple super single tyres	6	24	–	22.5

Traditionally, dual tyres have been used to limit pavement damage by efficiently distributing the axle loads over a larger contact area than single tyres, hence reducing the contact stresses on the pavement. Due to economic, safety and other benefits, the wide-base single (super single) tyres are increasingly used in the trucking industry. However, research studies demonstrate that wide-base single tyres cause more damage to pavements than do traditional dual tyres (Al-Qadi and Wang 2009; Greene et al. 2009; Abu Abdo 2017).

3 Study Approach

3.1 Description of the Case Study

The road section considered in this study, an approximately 40 km-long single carriageway with one lane in each direction, is located between Korogwe and Mombo towns in Tanzania's Tanga region. The Korogwe-Mombo road section forms part of the North-East corridor (T2 trunk road), which is the main trunk road that connects Tanzania's East Coast (including Dar es Salaam city and Tanga Port) with the northern regional cities of Kilimanjaro and Arusha. The T2 trunk road is also the main road that links Dar es Salaam and Nairobi, the major trade centres of Tanzania and Kenya respectively. In addition, the road is used by heavy vehicles travelling to and from the neighbouring country of Uganda.

3.2 Traffic Counts

The first traffic counts that were carried out in 2005 along the Korogwe-Mombo road included classified manual traffic counts. The classified manual traffic counts were carried out over seven consecutive days continuously for 12 h for the first four days, followed by 24-hour counts for the next three days. Vehicles passing the counting point were recorded separately for each direction. The traffic counts data was used for the structural design of the pavement for rehabilitation of the road section in 2006 (TANROADS 2006).

The second set of traffic counts was carried out in 2015 for the same Korogwe-Mombo road section after the rehabilitation of the section had been completed and the road had been opened to traffic (O'Connell et al. 2016). The data set consisted of manual classified traffic counts conducted over seven consecutive days for 24 h. During the traffic counts for both 2005 and 2015, the heavy vehicles were grouped into four categories (MGVs, HGVs, VHGVs and Buses), as recommended in the Tanzania Field Testing Manual (TANROADS 2003).

For pavement design and analysis purposes, heavy vehicle traffic count data is usually expressed in terms of the Annual Average Daily Traffic (AADT). The current study analyzed traffic count to determine AADT, and Eq. (1) used the AADT values based on 2005 traffic counts to compute the expected future traffic volumes in 2015. The traffic growth rates used for the computation of future traffic were 7.0% for Buses and 6.0% for MGVs, HGVs and VHGVs. These rates were estimated based on the recommendations contained in Tanzania's traffic growth baseline survey report (TANROADS 2009), and they were similar to those used during the rehabilitation design of the Korogwe-Mombo road.

$$\text{Projected AADT} = \text{Initial AADT} \times (1 + 0.01 \times j)^n \quad (1)$$

where j is traffic growth rate (%) and n is the time in years between the determination of traffic volume and the projection year.

3.3 Axle Load Surveys

Similar to the traffic counts, two sets of axle load measurements were carried out in 2005 and 2015 respectively. Both axle load measurements were conducted over seven consecutive days for 24 h, using a portable static weigh pad. The axle load surveys grouped the heavy vehicles into four categories: MGVs, HGVs, VHGVs and Buses. In addition to the axle load measurements, an Origin-Destination (OD) survey was also performed during the 2015 axle loads survey, including the type of the load/goods transported.

An Axle Equivalency Factor (AEF) is generally used to process the axle load survey data and assist with estimating the traffic loading for pavement design and analysis. The AEF represents the damaging effect of an axle passing over the pavement and is calculated using Eq. (2), as recommended in the Tanzania Field Testing Manual (TANROADS 2003).

$$\text{Axle Equivalency Factor (AEF)} = \left[\frac{\text{Axle Load (kg)}}{8160}\right]^{4.5} \tag{2}$$

For most pavement design purposes, the full axle load distribution is usually not available. As such, the concept of "E80 per heavy vehicle" (E80/HV) or Vehicle Equivalency Factor (VEF) is used. VEF is a factor that converts different truck loads to an equivalent number of standard axles (i.e. 8160 kg per axle). Equation 3 was used to process the axle load survey data of the individual heavy vehicles and determine VEF.

$$\text{Vehicle Equivalency Factor (VEF)} = \sum_{2}^{i} \text{AEF} \tag{3}$$

In Eq. (3), i is the total number of axles, and *AEF* is the Axle Equivalency Factor that was computed using Eq. (2). The VEF values of each heavy vehicle were subsequently used to determine the average VEF for each of the heavy vehicle categories. The average VEF was calculated separately for each direction/lane.

4 Analysis Results and Discussions

4.1 Assessment of Heavy Vehicles Volume

As mentioned earlier, for pavement design and analysis purposes, heavy vehicle traffic counts data are usually expressed in terms of the AADT. Figure 3 presents the AADT for each of the heavy vehicle categories for traffic counts carried out in 2005 and 2015. As theoretically expected, the AADT values for the 2015 traffic count are higher than those for 2005. For both 2005 and 2015 traffic count surveys, the AADT in both directions appears to be similar, with the exception of the MGV category for the 2015 survey (i.e. AADT for Korogwe to Mombo direction is 94, whereas the AADT for the opposite direction is 112). The traffic count data also shows that in 2005 the AADT for Buses was highest followed by VHGV, MGV and HGV categories. In contrast, the 2015 traffic count data indicates that the AADT for VHGV was the highest followed by Buses, MGVs and HGVs.

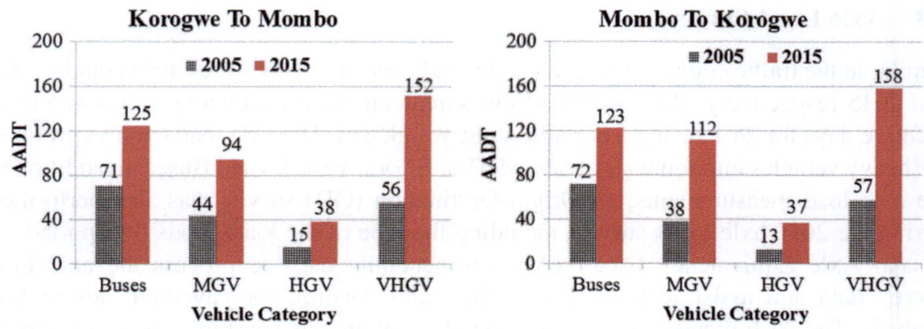

Fig. 3. AADT for 2005 and 2015 traffic counts

Figure 4 compares the actual traffic volumes obtained during the 2015 traffic counts and the volumes projected based on the 2005 traffic counts (using Eq. 1). With the exception of Buses, the AADT projections based on the 2005 traffic count data are generally lower than the actual AADT determined on the basis of the actual 2015 survey. For instance, while the projected AADT for Buses is approximately 12% higher than the actual AADT in the Korogwe-Mombo direction, the projected AADT for MGVs, HGVs and VHGVs is lower than the actual AADT by approximately 16, 29 and 34% respectively. The difference is significantly higher for the VHGVs category. The Origin-Destination (OD) survey performed as part of the axle loads survey indicated that most of the VHGV transport cement from the Tanga cement factory to the northern regions of Kilimanjaro and Arusha, and hence they may not have been accounted for during the 2005 traffic counts. Additionally, the general traffic growth trends over a 10-year period could be a contributing factor.

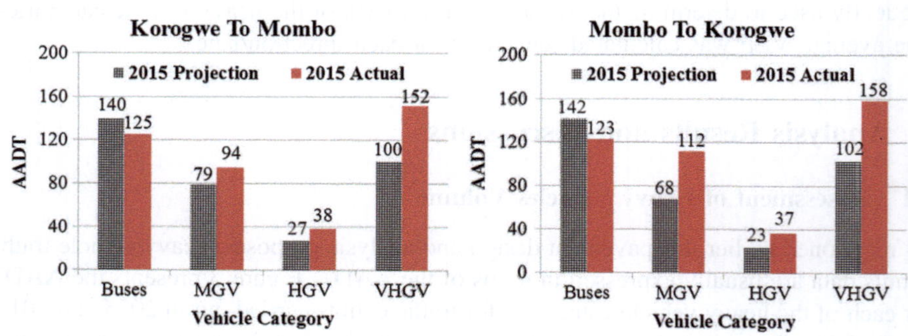

Fig. 4. Comparison of actual and projected 2015 AADT

The comparison of the actual and projected traffic suggests that the generic traffic growth rates obtained from the Tanzanian traffic growth baseline survey report should be used cautiously, as they may not be realistic for some roads. It further demonstrates the need to accurately determine site-specific traffic data for pavement structural design purposes, as opposed to using generic traffic growth rates generally derived from the expected growth in Gross Domestic Product (GDP) of a country. It should, however, be mentioned that both the 2015 and the 2005 traffic counts were carried out over a short period (seven days), hence the effects of seasonal variation of traffic may not have been accounted for.

4.2 Assessment of Axle Loads

Figure 5 compares the average VEF for each of the heavy vehicle categories that were determined using the 2005 and 2015 axle load surveys. For both 2005 and 2015 surveys, the VEF values determined for the Korogwe-Mombo direction are higher than those for the opposite direction (i.e. Mombo-Korogwe). This observation is in agreement with the Origin-Destination (O-D) survey data, which indicated that the heavy vehicles travelling in the Korogwe-Mombo direction are loaded more heavily than those travelling in the Mombo-Korogwe direction. The O-D survey indicated that the most common loads/goods transported by the heavy vehicles were cement (mostly from the Tanga cement factory), diesel/petrol, shop supplies, wheat flour, fertilizer, gas, building materials and farm produce to the northern regions of Tanzania and neighbouring countries of Kenya and Uganda. This observation demonstrates the importance of undertaking axle load surveys for each road direction separately, as traffic loading in opposite directions may differ significantly. It is further observed that the VEF values determined from the 2015 surveys are higher than those determined from the 2005 survey. Hence the VEF determined from the 2005 axle load survey data is more likely to underestimate the cumulative pavement traffic loading, as will be shown in the next section.

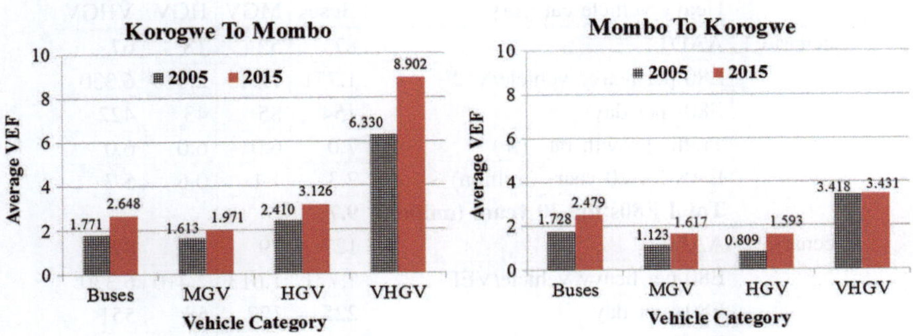

Fig. 5. Average vehicle equivalency factors (VEFs)

4.3 Comparison of Traffic Loading

The determined VEF values for the Korogwe-Mombo direction were used in combination with the AADT to determine the cumulative pavement loading (E80s). Because the values were higher than in the Mombo-Korogwe direction, they were critical for pavement design (i.e. the heavily loaded direction). A 20-year design period was assumed, which is the same as the design period used during the rehabilitation design of the road section. The commutative pavement traffic loading was computed using traffic counts and axle load survey data for both 2005 and 2015. As indicated earlier, the traffic growth rates used were 7.0% for Buses and 6.0% for MGVs, HGVs and VHGVs. The following three different scenarios were considered:

- Scenario 1: Use of the traffic counts and axle load survey data for 2005 to determine the 20-year traffic loading, with 2008 as the base year. This is similar to the approach used during the design of the rehabilitated road (i.e., the construction was originally planned to be completed by 2008).
- Scenario 2: Use of traffic counts and axle load survey data obtained in 2005, with 2015 as a base year. This means that the recommended traffic growth rates were applied to the actual 2005 traffic counts to project AADT for the year 2015 and then compute the 20-year traffic loading.
- Scenario 3: Use of the actual traffic counts and axle load survey data for the 2015 survey to determine 20-year traffic loading.

Tables 3 presents the calculated 20-year cumulative pavement traffic loading (E80s) for Scenarios 1 to 3 above as 9.7, 14.8 and 27.2 million respectively. As theoretically expected, due partly to traffic growth over a 10-year period, the cumulative traffic loading that was calculated based on the 2015 data (Scenario 3) is significantly higher than the estimated traffic loading based on the 2005 survey data using Scenarios 1 and 2 (i.e. approximately 2.8 times and 1.5 times higher compared with Scenarios 1 and 2 respectively). This was expected due to the higher AADT and VEF

Table 3. Cumulative traffic loading for Korogwe-Mombo direction

	Heavy vehicle category	Buses	MGV	HGV	VHGV
Scenario 1	AADT	87	52	18	67
	E80 per heavy vehicle/VEF	1.771	1.613	2.410	6.330
	E80s per day	154	85	43	422
	Traffic growth rate (%)	7.0	6.0	6.0	6.0
	E80s for 20 years (million)	2.3	1.1	0.6	5.7
	Total E80s for 20 years (million)	**9.7**			
Scenario 2	AADT	127	79	24	87
	E80 per heavy vehicle/VEF	1.771	1.613	2.410	6.330
	E80s per day	225	127	58	551
	Traffic growth rate (%)	7.0	6.0	6.0	6.0
	E80s for 20 years (million)	3.4	1.7	0.8	7.4
	Total E80s for 20 years (million)	**14.8**			
Scenario 3	AADT	125	94	38	152
	E80 per heavy vehicle/VEF	2.648	1.971	3.126	8.902
	E80s per day	331	185	119	1353
	Traffic growth rate (%)	7.0	6.0	6.0	6.0
	E80s for 20 years (million)	5.0	2.5	1.6	18.2
	Total E80s for 20 years (million)	**27.2**			

values obtained during the 2015 traffic counts and axle load survey respectively. Thus, the traffic loading that was computed based on the 2005 survey data may underestimate the actual expected future traffic loading, and it illustrates the importance of using the latest, most accurate and reliable traffic data during rehabilitation design.

4.4 Implications for Structural Pavement Design

According to the Tanzania Pavement Design Manual (1999), the traffic loading that was computed based on the 2005 traffic counts and axle load survey data falls under Traffic Loading Class-TLC 10 (i.e. cumulative traffic loading between 3 and 10 million E80s) for scenario 1 and TLC 20 (between 10 and 20 million E80s) for Scenario 2. On the other hand, the outcomes of the latest traffic and axle load surveys (conducted in 2015 after construction completion and the opening of the road to traffic) indicate that the road section is more likely to carry traffic loading that is equivalent to TLC 50 (between 20 and 50 million E80s) over the 20-year design life. The structural design for the rehabilitation of the road section was undertaken using TLC 10, which may have underestimated the expected future traffic loading.

Although a detailed evaluation of the structural adequacy of the pavement structure used for the rehabilitation of the Korogwe-Mombo road section falls outside the scope of this paper, the traffic loading analyses indicate that the pavement structure may have been under-designed and may require further structural strengthening before the end of the desired 20-year service life. However, monitoring the long-term performance of the road section may be needed to ascertain the extent to which an inaccurate determination of the traffic loading may shorten the pavement service life. It is also important to emphasize that the performance of pavements is not only dependent on the accurate determination of the expected future traffic loading. Other factors such as the quality of construction, supervision, design, materials, and climatic conditions may also affect the expected performance of the pavement.

5 Conclusions and Recommendations

By using the Korogwe-Mombo road section in Tanzania as a case study, this paper presented a comparative assessment of the traffic loading estimated during the rehabilitation design, the currently measured site-specific traffic loading, and the projected future traffic loading. Based on the results and discussions contained in this paper, the following conclusions are drawn and recommendations made:

- The assessments that were conducted demonstrated the need for good quality and reliable up-to-date traffic data to ensure an accurate determination of traffic loading for pavement design purposes.
- The common practice to conduct traffic studies over a short period of time (usually seven days) may cause significant errors in the prediction of design traffic loading due to the inability to capture seasonal variation of traffic; traffic pattern changes resulting from short- to medium-term changes in economic policies; transportation regulations; legal axle load limits; the level of enforcements; etc.

- The generic traffic growth rates used to determine traffic loading should be applied cautiously. Traffic volumes may vary over time due to, for instance, fluctuation in economic trends. Furthermore, the growth rates should not be assumed to be the same for different heavy vehicle categories, as traffic patterns may fluctuate for a specific heavy vehicle category.
- To improve the accurate determination of traffic loading, traffic studies should be conducted over long periods. Road agencies should also consider investing in the installation of permanent automated traffic and WIM-monitoring systems.
- In situations where significant time delays are expected from the rehabilitation design up to the start of construction, it is recommended that updated traffic studies be conducted to ascertain the traffic figures, as short- to medium-term changes in economic activities may affect traffic patterns (such as the cement factory influencing one direction in this paper).

Overall, this study has demonstrated the importance of periodic traffic surveys to measure and accurately quantify the changes and growth trends in traffic patterns, both in terms of volume counts and axle loads. In lieu of manual traffic surveys, use of more accurate automated traffic and WIM-monitoring systems is strongly recommended to ensure accurate traffic data characterization for optimal pavement design, rehabilitation and planning purposes. While costly permanent traffic and WIM-monitoring systems can be stationed on selected highway sites, portable WIM technology offers a cost effective alternative for deployment and traffic data measurements on any desired highway location.

Acknowledgements. The authors would like to acknowledge the Tanzania National Roads Agency (TANROADS) and the South African Council for Scientific and Industrial Research (CSIR) for funding this study.

References

Abu Abdo, A.M.: Effect of dual versus super single truck tyre on flexible pavement performance: a mechanistic approach. ARPN J. Eng. Appl. Sci. **12**(13), 4136–4141 (2017)

Al-Qadi, I.L., Wang, H.: Evaluation of pavement damage due to new tyre designs. Research Report ICT-09-048, Illinois Centre for Transportation, IL USA (2009)

De Beer, M., Fisher, C., Jooste, F.J.: Determination of pneumatic tyre/pavement interface contact stresses under moving loads and some effects on pavements with thin asphalt surfacing layers. In: Proceedings of the 8th International Conference on Asphalt Pavements, University of Washington, USA (1997)

Federal Highway Administration (FHWA): Traffic Monitoring Guide. Washington D.C. (2013)

Government Notice No. 30.: The Road Traffic (Maximum Weight of Vehicles). United Republic of Tanzania (2001)

Government Notice R225: The National Road Traffic Regulation: Part IV. Republic of South Africa, Loads on Vehicles (2000)

Greene, J., Toros, U., Kim, S., Byron, T., Choubane, B.: Impact of wide-base single tyres on pavement damage. Research Report FL/DOT/SMO/09-528, State Materials Office, Florida, USA (2009)

MOW: Pavement and Materials Design Manual. Ministry of Works, Dar es Salaam, Tanzania (1999)

O'Connell, J., Mgangira, M., Komba, J., Mturi, G.: Investigating the cause of deformation along the T2 Route between Korogwe and Mkumbara. CSIR Report Number CSIR/BE/IE/ER/2015/0067/B. Pretoria, South Africa (2016)

TANROADS: Field Testing Manual. Ministry of Works, Dar es Salaam, Tanzania (2003)

TANROADS: Detailed design for the rehabilitation of Korogwe-Mkumbara-Same Road. Design report: volume 1. Prepared by SMEC International, Australia in association with Engineering Research Associated (ERA), Tanzania (2006)

TANROADS: Consultancy services for carrying out baseline traffic counts in Tanzania mainland and establishing a comprehensive traffic census methodology for TANROADS. Prepared by Intercontinental Consultants and Technocrats Pvt. Ltd., India in association with Data Consult Limited, Tanzania (2009)

TMH 3: Specifications for the Provision of Traffic and Weigh-in-Motion Monitoring Service. Committee of Transport Officials, Pretoria, South Africa (2015)

TMH 14: Traffic and Axle Load Monitoring Procedures. Committee of Transport Officials, Pretoria, South Africa (2014)

TRH 4: Structural Design of Flexible Pavements for Interurban and Rural Roads. Department of Transport, Pretoria, South Africa (1996)

TRH 16: Traffic Loading for Pavement and Rehabilitation Design. Department of Transport, Pretoria, South Africa (1991)

TRL: Overseas Road Note 40: A Guide to Axle Surveys and Traffic Counts for Determining Traffic Loading on Pavements. TRL Limited, Crowthorne, Berkshire (2004)

Effect of Width of Geosynthetic Reinforcement within the Granular Cover on the Load Distribution over the Tunnel Lining

Yan Kou[1(✉)], Sanjay Kumar Shukla[2], and Alireza Mohyeddin[2]

[1] School of Engineering, Edith Cowan University, Joondalup, Perth, WA 6027, Australia
ykou@our.ecu.edu.au, kouyan_1989@hotmail.com
[2] Discipline of Civil and Environmental Engineering, Edith Cowan University, Joondalup, Perth, WA 6027, Australia
{s.shukla,a.mohyeddin}@ecu.edu.au

Abstract. A realistic estimation of load distribution over the buried structures is necessary for proper analysis of tunnels, culverts and pipes/conduits. Tunnels with linings are often constructed in transportation and hydraulic engineering. For the design of tunnel lining, it is essential to know the load over the lining. Load distribution over the buried structures has been investigated scientifically during the past several decades. The method of investigation includes experimental, numerical and analytical methods. The finite-element models based on some commercial software have been developed for load analyses for design of the tunnel linings and buried structures. The geosynthetic is an effective reinforcement layer to reduce the load over the buried structure. Although some studies have indicated that the geosynthetic layer can reduce the load over the buried structure, but no attempt has been made to determine the optimal width of the geosynthetic reinforcement within the granular cover. Therefore, in this paper, an attempt is made to present effect of width of geosynthetic layer on the load distribution over the tunnel lining. The study has been carried out by developing a numerical model of the problem. The commercial software PLAXIS 2D has been used for numerical modelling. The results have been presented in the form of design charts, mentioning the optimum width of geosynthetic layer, so that they can be used by practising engineers.

Keywords: Tunnel lining · Granular soil cover · Load distribution
Numerical modelling · Geosynthetic reinforcement

1 Introduction

Buried structures have been studied for centuries as tunnel linings, pipes and culverts. Tunnels were first constructed for underground mining in Egypt, Austria and some other countries around 3000 B.C. Tunnels for water supply started from A.D. 50, and for road, military and buried purposes after around A.D. 600. Civil excavations became popular around 19th and 20th centuries (Tatiya 2005). Many tunnelling projects have

© Springer International Publishing AG, part of Springer Nature 2019
A. Sevi et al. (eds.), *Enhancements in Applied Geomechanics, Mining, and Excavation Simulation and Analysis,* Sustainable Civil Infrastructures,
https://doi.org/10.1007/978-3-319-95645-9_6

been undertaken such as waterways, subways, railways and highways. In the current accelerated society, tunnels play a very important role in the transport purpose.

For the safety aspect, the relationship between tunnels and surrounding soil is an important factor to expect the failure in the future. That is why many researchers have investigated the load distribution on the buried structures, such as Valsangkar and Britto (1978, 1979), McVay and Pappadopoulos (1986), McVay et al. (1993), Li (2016), and Talesnick et al. (2008).

One of the reinforcements for buried structures is geosynthetic reinforcement, which is placed above the tunnel lining or buried pipe. Geosynthetics is a generic name representing a broad range of planer products manufactured from polymeric materials. Several types of geosynthetics are often used in the real field applications, such as geotextiles, geogrids, geonets, geomembranes and geocomposites, which are used in contact with soil, rock and/or any other civil engineering-related material as an integral part of a man-made project, structure or system (Shukla and Yin 2006). In buried structures, the effects of geosynthetic layer have been studied by some researchers, such as Dancygier and Yankelevsky (1996), Kawabata et al. (2003), Ahmed et al. (2015), and Shukla and Sivakugan (2013). However, these works did not deal with the effect of width of geosynthetic layer over the tunnel lining.

In this paper, the load reductions by using geosynthetic reinforcement within the granular cover over the tunnel lining with surface loading are presented numerically to supplement the very limited information that exists for tunnelling applications. In addition, the width of geosynthetic layer is varied to investigate its effect on the load distribution over the tunnel lining, and graphical presentations are made for their applications as design charts.

1.1 Numerical Formulation

A series of finite-element analyses of tunnel lining covered with geosynthetic reinforcement was conducted using the PLAXIS 2D commercial software. The 15-node triangle elements were used to model the soil mass. Mohr-Coulomb failure criterion was used for the simulation of the soil behaviour. This failure criterion is a well-known linear-elastic, perfectly plastic model. It has a constant average stiffness estimation for the soil layer. The soil properties of model are given in Table 1.

A sensitivity analysis was carried out to identify the minimum distance to place the boundaries for the geometry and loading conditions to minimize the boundary effects. The determination from sensitivity analysis was found to be 8 times the diameter of the tunnel (Bryden et al. 2015). Therefore, the numerical test box was selected with dimensions as 40 m × 40 m. AB = 6 m wide strip of the ground was displaced by 80 mm, keeping the strip centreline coinciding with the tunnel, as shown in Fig. 1.

Table 1. Property of sand used in numerical model

Property	Value
γ_{unsat} (kN/m^3)	17
γ_{sat} (kN/m^3)	20
E (kPa)	1.3×10^4
v	0.3
C_{ref} (kPa)	1
ϕ (degree)	30
ψ (degree)	0

Note γ_{unsat} = unsaturated unit weight; γ_{sat} = saturated unit weight; E = effective Young's modulus; V = effective Poisson's ratio; C_{ref} = *effective cohesion;* ϕ = effective friction angle; ψ = diliatancy angle

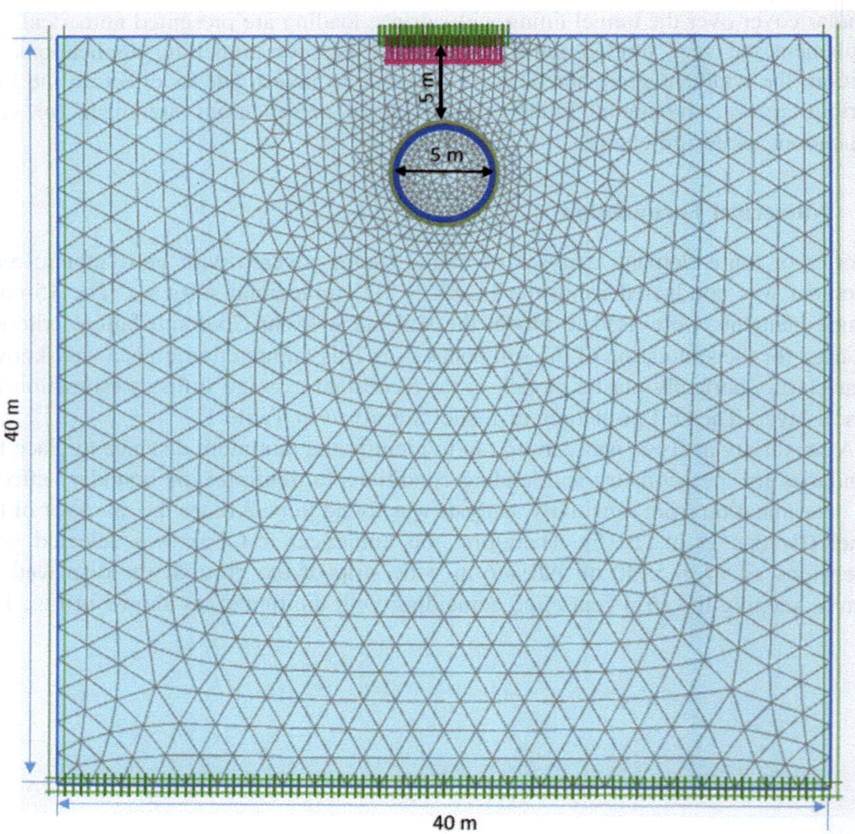

Fig. 1. Mesh generation for the tunnel covered with unreinforced soil

The tunnel lining was modelled using the plate element function with elastic behaviour as assumed. The axial stiffness *EA* and flexural rigidity *EI* based on the literature were assumed to be 1.4×10^7 kN/m and 1.43×10^5 kN m^2/m, respectively with Poisson's ratio, $\nu = 0.15$. The crown of the tunnel lining, having diameter, $B_c = 5$ m, was placed 5 m below the ground surface. Interface elements were used to simulate the interaction between soil and tunnel lining, and soil and geogrid.

Geogrid is one of the geosynthetic reinforcements. It was placed at a depth of 4 m below the surface, that is, 1 m above the tunnel lining. BX1100 type geogrid having a tensile modulus of 205 kN/m was used. Its width b was varied from B_c to $6B_c$. The centreline of the geogrid was always over the centreline of the tunnel lining, as shown in Fig. 2.

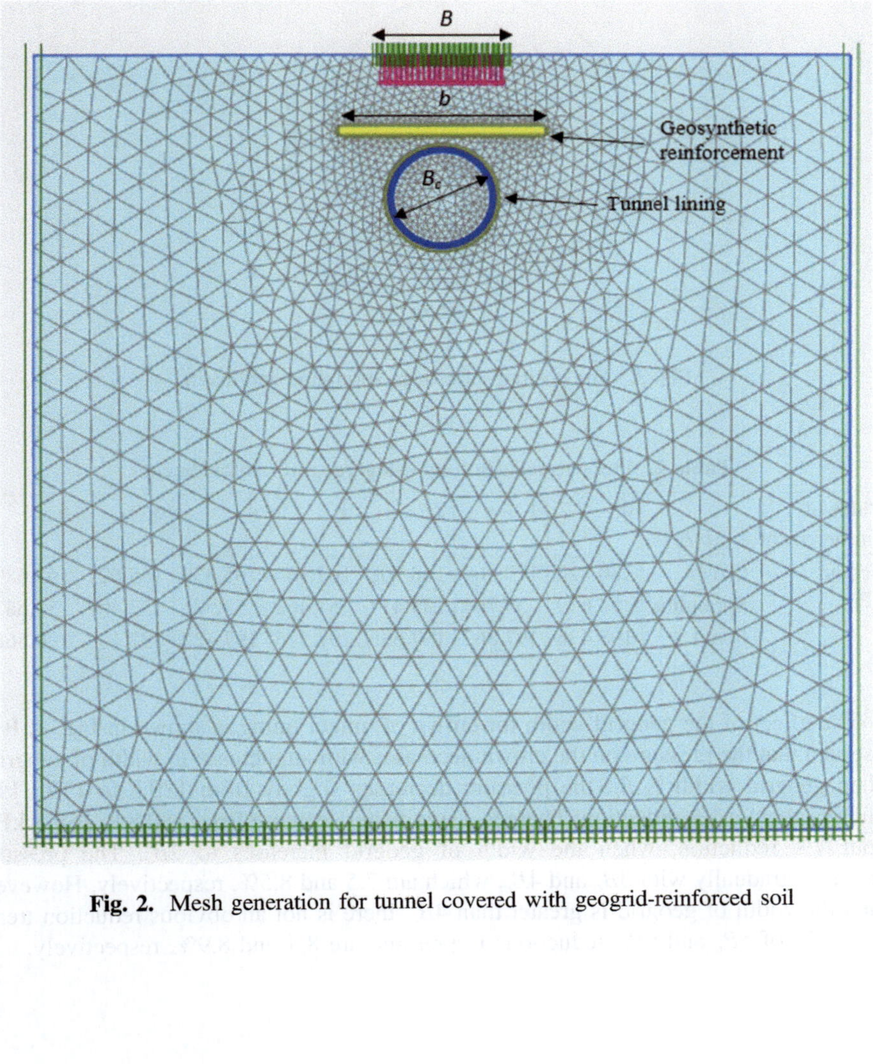

Fig. 2. Mesh generation for tunnel covered with geogrid-reinforced soil

1.2 Results and Discussion

The numerical model as explained in the previous section was developed to investigate the effect of width of geosynthetic reinforcement over the tunnel lining. Three points around the tunnel lining were analysed: crown (top) of tunnel lining, springline (right) of tunnel lining and invert (bottom) of tunnel lining, as shown in Fig. 3. Six different widths of geogrid have been investigated which range between diameter of the tunnel lining (B_c) to six times diameter of the tunnel lining ($6B_c$). The load distribution on those three points for both unreinforced and reinforced cases are presented in Table 2.

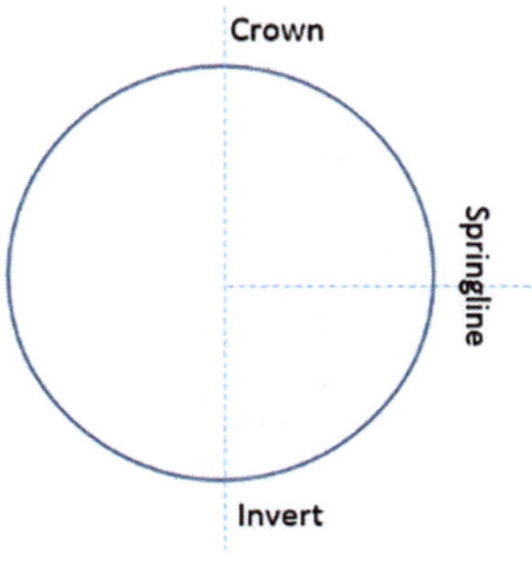

Fig. 3. The position of crown, springline and invert

Table 2. Pressure reduction by using geogrid reinforcement

Width of reinforcement → (b/B_c)		0	1	2	3	4	5	6
Pressure (kPa)	Crown	180.822	179.064	168.508	167.167	165.336	165.621	164.689
	Springline	54.619	53.764	53.417	53.462	53.501	53.407	52.682
	Invert	198.436	197.862	197.636	197.575	197.422	197.16	196.568

The effect of the geogrid width for crown of tunnel lining is shown in Fig. 4. It is observed that the pressure on the crown decreases with an increase in width of geogrid. When geogrid width is B_c, the pressure decreases slightly from 180.8 to 179.1 kPa which is about a 1% reduction. There is a significant decrease from 180.8 to 168.5 kPa, about 7% reduction, when the width of geogrid increases to $2B_c$. The pressure decreases gradually with $3B_c$ and $4B_c$, which are 7.5 and 8.5%, respectively. However, when the width of geogrid is greater than $4B_c$, there is not an obvious reduction trend for widths of $5B_c$ and $6B_c$, reductions in pressure are 8.4 and 8.9%, respectively.

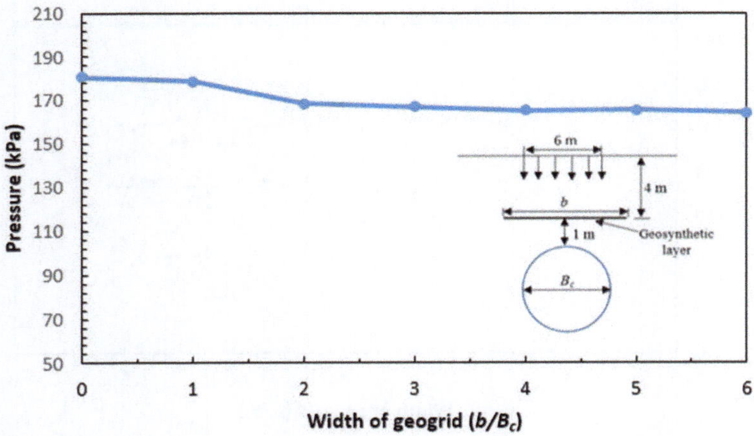

Fig. 4. Variation of pressure on the crown of the tunnel lining with width of geogrid reinforcement (b/B_c)

Figure 5 shows the load distribution on the springline with different widths of geogrid. It is noticed that the pressure decreases with an increase in width of geogrid, but the pressure reduction was not very prominent. For example, the pressure is 54.6 kPa for unreinforced case while it is to 52.7 kPa with a geogrid width of $6B_c$, thus only a reduction of about 3.6% takes place.

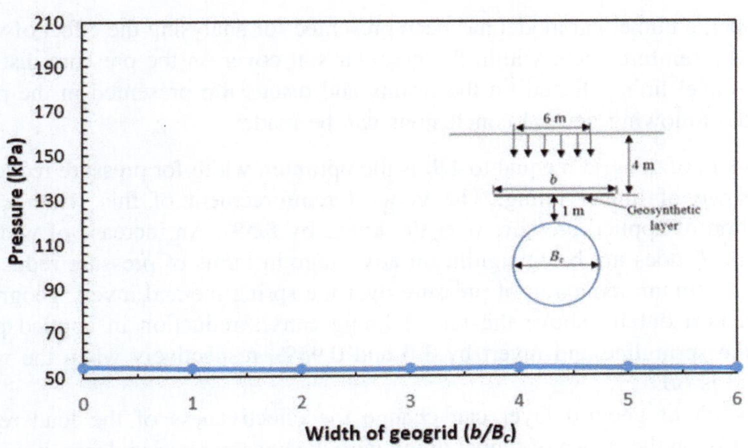

Fig. 5. Variation pressure on the springline of the tunnel lining with width of geogrid reinforcement (b/B_c)

Figure 6 presents how the pressure on the invert varies with different widths of geogrid. There is a decrease of the pressure on the invert when the width of the geogrid increases from B_c to $5B_c$; however overall, it is relatively ineffective in reducing the

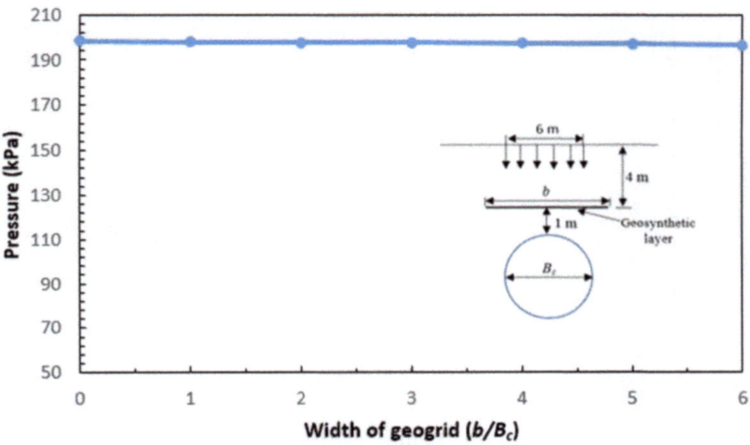

Fig. 6. Variation of pressure on the invert of the tunnel lining with width of geogrid reinforcement (b/B_c)

pressure on the invert of the tunnel lining. Compared to the unreinforced situation, even when the width of the geogrid is $6B_c$, it only decreases when the loading pressure is increased from 198.4 to 196.6 kPa, resulting in roughly a decrease by 0.95%.

2 Conclusions

In this paper, a numerical model has been presented for analysing the effect of width of geosynthetic reinforcement within the granular soil cover on the pressure distribution over the tunnel lining. Based on the results and discussion presented in the previous section, the following general conclusions can be made:

1. The width of geogrid b equal to $4B_c$ is the optimum width for pressure reduction on the crown of tunnel lining. The geogrid reinforcement of this width causes a reduction of applied pressure over the lining by 8.5%. An increase of width from $4B_c$ to $6B_c$ does not bring significant advantage in terms of pressure reduction.
2. For a maximum reduction of pressure over the springline and invert, geogrid layer placed horizontally above the tunnel lining causes reduction in applied pressure over the springline and invert by 3.6 and 0.95%, respectively when the width of geogrid is $6B_c$.
3. The width of geogrid layer can change the effectiveness of the load reduction obviously on the crown of tunnel lining with the constant buried depth, however the buried depth of the geogrid is another factor that may affect the load reduction. Hence, future studies may take place on investigating how the buried depth of the geogrid layer above the tunnel lining affects the pressure reduction.

Reference

Ahmed, M.R., Tran, V.D.H., Meguid, M.A.: On the role of geogrid reinforcement in reducing earth pressure on the buried pipes: experiment and numerical investigations. Soil Found. **55** (3), 588–599 (2015)

Bryden, P., Naggar, H.E.I., Valsangkar, A.: Soil-structure interaction of very flexible pipes: centrifuge and numerical investigations. Int. J. Geomech. (2015). https://doi.org/10.1061/(ASCE)GM.1943-5622.0000442

Dancygier, A.N., Yankelevsky, D.Z.: A soft layer to control soil arching above a buried structure. Eng. Struct. **18**(5), 378–386 (1996)

Kawabata, T., Uchida, K., Hirai, T., Mohri, Y., Ling, H. I., Koyama, N.: Experiment on buried pipe using backfill of cover with geosynthetics. In: New Pipeline Technologies, Security, and Safety Conference; Pipeline 2003, ASCE (2003)

Li, C.: Developing an analytical method to study vertical stress due to soil arching during tunnel construction. J. Geotech. Geol. Eng. **34**, 1247–1255 (2016)

McVay, M.C., Pappadopoulos, P., Bloodmquist, D.: Long-term behaviour of buried large-span culverts in cohesive soil. Transportation Research Record 1414, Transportation Research Board, Washington, DC, pp. 40–46 (1993)

McVay, M.C., Pappadopoulos, P.: Long-term behaviour of buried large-span culverts. J. Geotech. Eng. **112**(4), 424–442 (1986)

Shukla, S.K., Sivakugan, N.: Load coefficient for ditch conduits covered with geosynthetic-reinforced granular backfill. Int. J. Geomech. **13**(1), 76–82 (2013)

Shukla, S.K., Yin, J.-H.: Analytical Expression for Geosynthetic Strain Due to Deflection. Geosynth. Int. **16**(5), 402–407 (2006)

Talesnick, M., Horany, H., Dancygier, A.N., Karinski, Y.S.: Measuring soil pressure on a buried model structure for the validation of quantitative frameworks. J. Geotech. Geoenviron. Eng. (2008). https://doi.org/10.1061/(ASCE)GM.1943-5622.0000442

Tatiya, R.: Civil Excavations and Tunnelling—A Practical Guide. Thomas Telford Ltd., London (2005)

Valsangkar, A.J., Britto, A.M.: Centrifuge tests of flexible circular pipes subjected to surface loading. Supplemental Rep. SR530, Transportation and Rock Research Laboratory, Crowthorne, UK (1979)

Valsangkar, A.J., Britto, A.M.: The validity of ring compression theory in the design of flexible buried pipes. Supplemental Rep. SR440, Transportation and Rock Research Laboratory, Crowthorne, UK (1978)

Durability Assessment of Pavement Foundation Materials Treated with a Polymeric-Based Additive

Romel Georgees[1,2](✉), Rayya Hassan[1], and Robert Evans[1]

[1] Swinburne University of Technology, Melbourne, Australia
rgeorgees@swin.edu.au
[2] Kirkuk University, Kirkuk, Iraq

Abstract. Strength and durability characteristics of granular and subgrade pavement materials have been shown to improve using polymer additives. Using such types of additives have shown to lower carbon footprints compared to traditional cementitious additives. This study reports the outcomes of a laboratory investigation to evaluate the use of a polymeric stabilization technique in improving engineering properties of pavement foundation materials. A synthetic polyacrylamide-based additive (PAM) has been used to stabilize three types of soils, commonly used in pavement construction. Repeated load triaxial tests were conducted to assess the stiffness and resistance to permanent deformation characteristics of PAM-treated soils. Simple capillary rise and abrasion tests have also been conducted to assess durability characteristics. The results revealed a significant overall increase in resilient modulus and a remarkable decrease in permanent deformation for the treated samples, depending on soil type. Further, the PAM used herein was shown to enhance the sealing capacity of the soils with the rise of water table, and dramatically improved the abrasion resistance for all the soils tested.

1 Introduction

Australia has approximately 800,000 km of road network, which is listed as the tenth most extensive network in the world (FactBook 2016). Of this network, 60% of these roads are unsealed. This large network is expensive to maintain and consumes a significant amount of high quality natural materials, which is unsustainable. Attempts have been conducted to reduce this expense and improve sustainability by using lower quality aggregates and improving their engineering properties through different stabilization techniques.

Currently pavement designers have many types of binders to use in soil stabilization. The choice to use the most suitable for the soil type is crucial to obtain enhanced engineering properties and economic performance. This has made cement the most common binder for soil stabilization due to its compatibility with almost all soil types and availability (Vorobieff and Wilmot 2001). However, the development of non-traditional stabilizers, such as polymers have gained greater attention as they demonstrate effectiveness in the field with regards to reducing permeability, increasing

© Springer International Publishing AG, part of Springer Nature 2019
A. Sevi et al. (eds.), *Enhancements in Applied Geomechanics, Mining, and Excavation Simulation and Analysis,* Sustainable Civil Infrastructures,
https://doi.org/10.1007/978-3-319-95645-9_7

durability, and non-time dependence of mixing and compaction, as well as better sustainability outcomes (Wilmot 1994; Andrews and Sharp 2010; Camarena 2013).

Polymeric-based additives have been used in the stabilization of unsealed pavements as well as granular select fills (i.e. working platforms over weak subgrades) and subgrades of sealed pavements to enhance performance properties. However, very few studies have been carried out to assess the feasibility of using PAM stabilized granular materials within the structural layers of the pavement (base and subbase). The aim of this study is to evaluate the suitability of using PAM-treated granular materials that are currently used in wearing courses of unsealed pavements in subbase layers of sealed low volume road pavements. To achieve the aim of this study, a laboratory experimental program has been undertaken to assess the changes in stiffness and durability characteristics of pavement materials when treated with an off the shelf synthetic polyacrylamide-based stabilizing additive. The tests performed include repeated load triaxial test (RLT), capillary rise test and abrasion test. The outcomes of this study will contribute to improving the knowledge regarding the behaviour of PAM-treated materials within a pavement structure and help promote the reliability of these sustainable materials for wider adoption by road authorities.

2 Materials and Methods

Three different soil samples from three different sites in Victoria/Australia were selected for this study, denoted as soil I, II and III. The soil samples were collected from the top 150 mm of an existing wearing courses of several unsealed roads that were being stabilized with PAM. For the three soil types, plasticity indices were determined following Australian standard AS 1289.3.3.1 (AS 2009) and their classifications were determined using the unified soil classification system (USCS) (ASTM 2011). The classification and geotechnical properties of the tested soils are summarized in Table 1.

The polymeric additive used in this study was a synthetic soluble anionic PAM, which is produced in a granulated form. This PAM is a non-toxic water-soluble material with a specific gravity of 0.8 and a pH value of 6.9 at 25 °C, and has a high molecular weight typically between 12 and 15 Mg/mole.

For sample preparation, the PAM amount used was 0.002% by dry weight of the soil, according to supplier's recommendation. The PAM was first mixed with water in a sealed container at a rate of 2 g per 5 l which resulted in a concentration that was twice the recommended. The concentration of this solution was adjusted when adding water to achieve moisture requirement of each soil sample tested. The soil-water mixture was then mixed in a mechanical mixer for 10–15 min, and the mixture was kept in plastic bags for 24 h in a storage room at a temperature of 20 °C ± 2 to allow for an even moisture distribution.

The maximum dry density (MDD) and optimum moisture content (OMC) were determined using the Australian modified proctor compaction test (AS 2003). The samples were compacted in five layers at a compaction effort of 45 blows per layer (i.e. 4961 kN m/m^3) for soil type I and 35 blows per layer (i.e. 3868 kN m/m^3) for soil types II and III. This was changed from the standard 25 blows per layer (i.e.

Table 1. Soil geotechnical properties of tested soils

Properties	Soil I	Soil II	Soil III
% Gravel (19.0–4.75 mm)	33.2	29.6	13.2
% Sand (2.36–0.075 mm)	56.8	46.4	36.8
% Fines (<0.075 mm)	10.0	24.0	50.0
% Clay	1.0	2.75	13.5
Liquid limit (%)	22.2	23.8	31.4
Plastic limit (%)	N/A	12.8	15.1
Plasticity index (%)	N/A	11	16.3
Specific gravity	2.78	2.51	2.65
Water absorption (coarse fraction), %	2.19	11.26	2.46
Optimum moisture content (%)	5.6	8.5	12.5
Max dry density (g/cm^3)	2.35	2.01	1.95
Soil classification, (USCS)	Poorly-graded sand with silt (SP-SM)	Clayey sand with gravel (SC)	Sandy clay (CL)

2703 kN m/m^3) as these higher compaction efforts were found to be the optimum for each soil type and were determined after a number of trials to ensure that laboratory performance of the PAM-treated samples was comparable with relevant observed field performance. Thereafter, this variation in compaction energy was applied to both treated and untreated samples. The reader is referred to a study conducted by Georgees et al. (2015) that explains the reason for changing the compaction effort.

The repeated load triaxial (RLT) test was conducted in accordance with the recommended Australian test procedure (Austroads 2000). The specimens were prepared in split moulds of 100 mm in diameter and 200 mm in height and were compacted in 8 layers using mechanical compaction at a compaction effort corresponding to the soil type. Two specimens per sample (treated and untreated) for each soil type were prepared. All specimens were compacted to a target density of 100% maximum dry density at optimum moisture content. After compaction, specimens were removed carefully from the split moulds and left to dry back to 50% of OMC. These specimens were then subjected to the RLT test and the values of their permanent strain and resilient moduli were determined at different stress conditions.

Permanent deformation determination characterizes the vertical permanent strain responses under different stress conditions to examine the stress dependent permanent strain per single test. Specimens were loaded with three different loading stages, each involved 10,000 cycles, at stress condition of specified deviator stresses of 250, 350 and 450 kPa and a constant 50 kPa confining stress (Austroads 2000). The loading type used is a trapezoidal pulse with a total stress period of 3 s, with 1 s loading, and rise and fall of 0.3 s.

The resilient modulus determination characterizes the vertical resilient strain responses over 66 stress conditions at 200 cycles per condition, and a combination of applied repeated vertical (deviatoric) and static lateral (confining) stresses ranging from 100 to 600 kPa and 20 to 150 kPa, respectively. The increments of the stresses and stress ratios were small to avoid early failure, which can occur at high stress ratios (Austroads 2000; AASHTO 2003). In all test conditions, none of the specimens underwent damage due to cyclic loading.

The abrasion resistance test was performed according to the test method proposed by Sampson (1988). Here, soil samples were sieved to collect only particles less than or equal to 4.75 mm in diameter. The reason for selecting the maximum aggregate size of 4.75 mm was to eliminate the possibility of intermittent plucking of the coarse particles by the brush, which eliminates a continual loss of mass with increased revolution (Jones 2007). Test samples were prepared according to the procedure outlined earlier for the modified proctor compaction test and using relevant optimum compaction efforts. The specimens were then left to dry back until reaching a constant mass. At least three specimens per sample (treated and untreated) were prepared. The cylindrical specimens were tested by subjecting 500 revolutions of the brush loaded to 2.2 kg on the side of the specimen, as shown in Fig. 1. This was carried out in two equal stages of 250 revolutions to reproduce the long term traffic wear action in the laboratory (Sampson 1988). The brushed specimens were then weighed and the loss of mass as a percentage of the original mass was recorded.

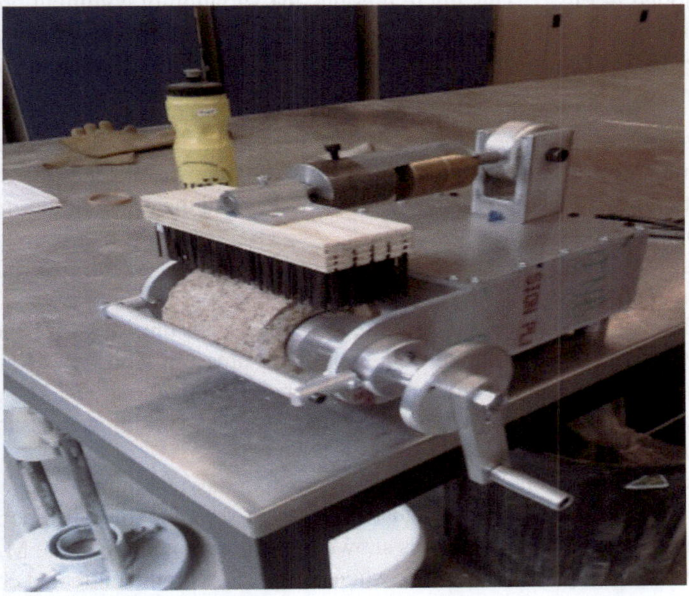

Fig. 1. Abrasion resistance apparatus

The capillary rise test was conducted on soil type II only and was performed according to Australian standard AS 1141.53 (AS 1996). Soil type II was chosen because of its high porosity compared to the other two soil types (i.e. soils I and III). The samples were prepared by mixing the soil samples in dry conditions with the pre-determined optimum water content. The samples were then kept in sealed plastic bags for 24 h in a storage room at a temperature of 20 °C ± 2 to allow even moisture distribution. The mixture was then placed in moulds (105 mm in diameter and 116 mm in height) and compacted in five layers using the relevant optimum compactive energy of 3868 kN m/m^3. The specimens were allowed to dry back to a constant mass, after which they were placed in water, at room temperature, in a dish to a depth of 10 mm. Capillary rise measurements were taken at different time intervals.

3 Results and Discussion

3.1 Stiffness and Permanent Deformation Characteristics

3.1.1 Resilient Modulus (M_R) Test Results

To successfully characterize flexible pavement materials using a mechanistic approach, resilient modulus values obtained from repeated load testing are required (AASHTO 1993; Austroads 2008). For each of the 66 stress stages of the RLT test, which consists of one value of confining stress and a corresponding deviatoric stress, the resilient modulus (M_R) values were determined for treated and untreated specimens for each soil type and plotted in Fig. 2. This figure presents the variation of confining pressures and the corresponding deviatoric stresses together with associated M_R values over the 66 stress stages of the RLT test. Careful examination of the figure resulted in the following observations:

- The resilient moduli for all three soil types are considerably affected by both deviatoric and confining stresses. Generally, M_R values over the 66 stress stages of all soil types increase with the deviatoric stresses, and their magnitudes are higher at high confining pressures.
- The addition of PAM to the soils enhanced their resilient moduli over all 66 stages of stress conditions
- The differences in M_R values of treated samples from untreated, for the three soil types, vary with the magnitude of the confining pressure. Soil types I and III show the greatest differences at high confining pressure (100 and 125 kPa), while the greatest difference for soil type II is found at lower confining pressures (20 and 75 kPa).
- When compared to the untreated counterparts, the maximum difference in the average M_R between untreated and PAM-treated samples for soil type I is 31.9%. Soil type III, on the other hand, has shown a maximum increase in the M_R values for the PAM-treated specimens at 46% when compared with the untreated counterparts. However, the maximum difference in M_R between treated and untreated samples for soil type II is only 8.8%.

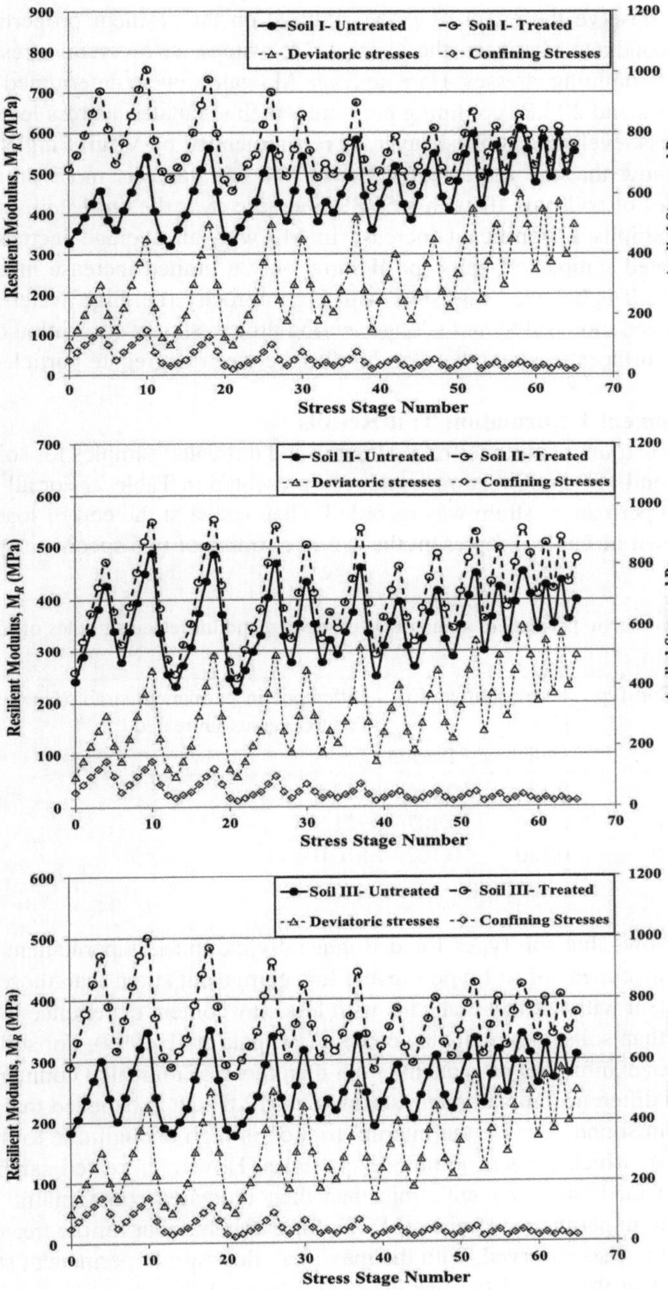

Fig. 2. RLT resilient modulus values of untreated and PAM-treated samples for soil types I, II and III at various stress conditions

To better observe the effect of PAM additive on the resilient properties, a comparison was conducted between the average M_R values, at a certain stress stage i.e. deviatoric and confining stresses. Here, average M_R values were determined at 100 kPa deviatoric stress and 20 kPa confining pressure which simulates a stress level closest to the in situ stress level for a subbase layer, as recommended by Virgil Ping et al. (2001). The results show that the impact of PAM on the M_R value is more pronounced in treated samples of soil type III with a 55.8% increase over the untreated samples. Soil type I also exhibits a significant increase in M_R with an average increase of 35%. However, treated samples of soil type II show only a limited increase in average M_R value of 8.8%. It is believed that soils with high porosity (i.e. high water absorption, see Table 1) need more PAM and a longer curing time to satisfy adsorption of the PAM onto the soil surfaces to strengthen the binding between aggregate particles.

3.1.2 Permanent Deformation Test Results

The permanent strain magnitudes of the treated and untreated samples for soil types I, II and III at the end of 30,000 loading cycles are presented in Table 2. For all specimens, the maximum permanent strain was recorded when tested at the end of loading cycles and those shown in Table 2 represent the average strains of two specimens per sample.

Table 2. Difference in permanent strain between treated and untreated samples of soil types I, II and III

Soil type	Permanent strain (%)		Difference in permanent strain (%) Treated versus untreated
	Untreated	Treated	
I	0.133	0.116	−12.8
II	0.136	0.107	−21.3
III	0.180	0.162	−10.0

Table 2 shows that soil types I and II generally exhibit less permanent strain than soil type III. Specimens of soil type I reveal less permanent strain than those of type II. This is consistent with the fact that soils with less clay content experience lower plastic deformations than soils with high clay content (Puppala et al. 1999). For soil type I, the treated samples exhibit less permanent strain than those of untreated counterparts, with the maximum difference in permanent strain being 12.8%. It is believed that the treated samples remain sound because the internal friction is high enough due to the bonding action of PAM, which prevent joining of particles. However, treated samples of soil type II exhibit far less permanent strain than their untreated counterparts. The maximum difference in permanent strain is 21.3%. Similar behaviour for the treated samples of soil type III is also observed, with the maximum decrease in permanent strain of the treated samples at the end of the loading cycles being 10%.

3.2 Durability Characteristics

3.2.1 Abrasion Resistance Test Results

The advantage of using PAM to stabilize pavement material is not limited to enhancing strength characteristics but also to provide a seal to the compacted subgrade and/or granular layer to resist abrasion due to construction traffic and also to resist water erosion in case of rainy condition. Figure 3 demonstrates the effect of PAM on increasing the abrasion resistance of the treated soils using the abrasion test. It is worth noting that the mass losses (in grams) for the samples (treated and untreated) presented in this figure are the average values of at least three specimens per sample.

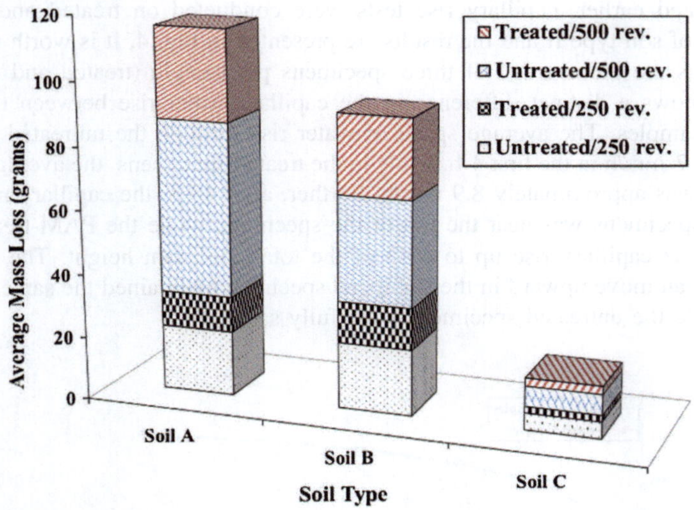

Fig. 3. Abrasion resistance test results for the treated and untreated samples of soils I, II and III

Figure 3 clearly shows that there is an increase in the mass loss for all soil types when the materials are subjected to higher brush revolutions. For the first 250 revolutions of abrasion, the mass loss is greatest for soil I and least for soil III. The mass loss is strongly dependent on the soil type (i.e. the amount of clay fines present in the soil); the abrasion resistance is highest for soil type III followed by soil type II and soil type I. This is supported by the role of clay in increasing the cohesion of materials. This trend is true for both treated and untreated soils, but less intensity was identified in the treated samples. Figure 3 also demonstrates higher abrasion resistance for the treated samples than the untreated samples. Abrasion resistance, at the end of 250 revolutions, is dramatically higher in the treated samples of soils I and III with an average of 45.1 and 60.9%, respectively than their untreated counterparts. Whereas soil II only showed a 38.1% increase in abrasion resistance when treated than untreated. On the other hand, increasing the brush load revolution to 500 did not change the trend and the mass losses for soils I, II and III were 48, 23.5 and 59.3%, respectively.

It is believed that the adsorption of the anionic PAM onto the clay particles occurs by cation bridging (Theng 1982), through which the polyvalent cations bridge the negative charged groups of clay particles and polymers together. On the other hand, bonding between particles is increased essentially by a dual process: First, the slipping action of PAM in the compaction stage, resulting in increased density and hence increased contact points per unit area between the soil particles; Second, the PAM molecules encapsulate the soil particles and upon drying an electrostatic attraction among soil particles is developed. These dual actions resulted in stronger bonds and less mass loss for the treated materials.

3.2.2 Capillary Rise Test Results

As mentioned earlier, capillary rise tests were conducted on treated and untreated specimens of soil type II and the results are presented in Fig. 4. It is worth noting that these results are the average of three specimens per sample (treated and untreated). Figure 4 shows a distinct difference in the capillary water rise between treated and untreated samples. The average speed of water rise through the untreated specimens reached 18.7 mm/h in the first 4 h, while in the treated specimens, the average speed of water rise was approximately 8.9 mm/h. Further, after 48 h, the capillary moisture of untreated specimens was near the top of the specimen, while the PAM-treated specimens had the capillary rise up to 40% of the total specimen height. The amount of water that can move upward in the stabilized specimen maintained the same level after 120 h, while the untreated specimens were fully saturated.

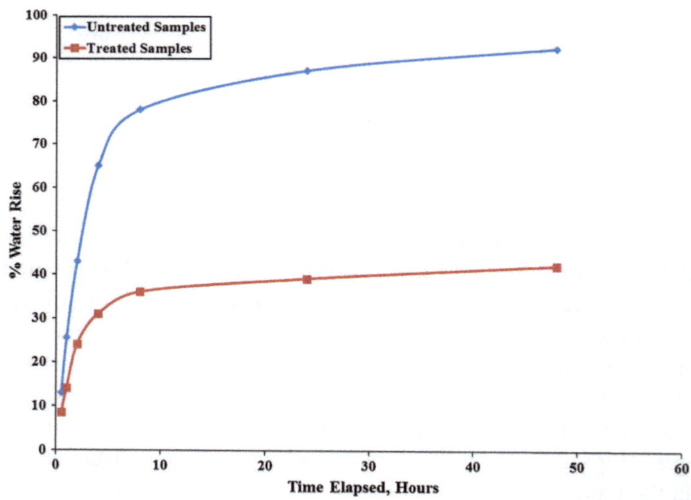

Fig. 4. Percentage water rise in treated and untreated samples of soil type II

As a matter of fact, capillary energy is a function of the liquid properties along with soil properties. In other words, if soil properties are kept constant, increased liquid wettability (e.g. contact angle with pore aggregate surface) and increased liquid surface tension will increase the capillary suction in the soils. However, in the case herein, when PAM is in contact with water, the solution's viscosity becomes higher, resulting in increasing the time of contact between water and soil particles (Malik et al. 1991), thereby the absorption time for the whole sample will increase. This action can be called "external waterproofing", rather than internal waterproofing.

4 Conclusions

This study focused on using polyacrylamide-based additive to improve the resilient responses and durability characteristics of pavement materials that can be used as a foundation materials for the sealed pavements. From the results of this study, the following has been concluded:

1. The addition of PAM increased the resilient modulus for the granular and cohesive soils tested by 31 and 46% respectively. However, the percentage increase in resilient modulus was not significant in Clayey Sand soil (soil II), which showed approximately 8.8% increase.
2. Soils treated with PAM had lower permanent strains than their untreated counterparts under repeated load triaxial test. The greatest difference in permanent strain between treated and untreated samples was found in the Clayey Sand soil, while the least difference was found in sandy Clay soil.
3. An increase in the resistance to abrasion was measured between 29.2 and 62.5% for all soil types treated with PAM. It was noted that the amount of fine clay in the soil was a factor in decreasing abrasion resistance.
4. Using PAM additive as a stabilizing agent has shown to reduce the capillarity action of soil. Samples treated with PAM showed capability of reducing the speed of rising water to 60% in the first 4 h of the test.

Overall, these results indicate that PAM is a viable alternative additive for sustainably improving the performance of pavement foundation materials with a potential of reducing pavement thickness. Its use in pavement foundation applications will also significantly lower the carbon footprint for future road infrastructure construction and maintenance.

Acknowledgements. The authors would like to acknowledge the logistic support of the Earthco projects company, particularly Mr. Mark Holding for providing the stabilizer additive for this research.

References

AASHTO: Guide for design of pavement structures. American Association of State Highway and Transportation Officials, AASHTO, Washington, D.C., USA (1993)

AASHTO: Standard method of test for determining the resilient modulus of soils and aggregate materials. American Association of State Highway and Transportation Officials, AASHTO, Washington, D.C. USA (2003)

Andrews, R., Sharp, K.: A protocol for conducting field trials for best value management of unsealed roads. In: 24th ARRB Conference—Building on 50 Years of Road and Transport Research (2010)

AS: Methods of sampling and testing aggregates Method 53: absorption, swell and capillary rise of compacted materials AS1141.53. Standards Australia, Australia (1996)

AS: Methods of testing soils for engineering purposes Method 5.2.1: soil compaction and density tests—determination of the dry density/moisture content relation of a soil using modified compactive effort. AS 1289.5.2.1. Standards Australia, Sydney, Australia (2003)

AS: Methods of testing soils for engineering purposes Method 3.3.1: soil classification tests—calculation of the plasticity index of a soil. AS 1289.3.3.1. Standards Australia, Sydney, Australia (2009)

ASTM: Standard practice for classification of soils for engineering purposes (unified soil classification system). ASTM D2487. West Conshohocken, PA, USA: American Society for Testing Materials (2011)

Austroads: Determination of permanent deformation and resilient modulus characteristics of unbound granular materials under drained conditions. Commentary to AG:PT/T053. Austroads, Sydney, NSW, Australia (2000)

Austroads: Guide to Pavement Technology Part 2: Pavement Structural Design. Sydney, NSW, Australia (2008)

Camarena, S.: Sustainable Road Maintenance and Construction Utilising New Technologies. International Public Works Conference, Darwin, Northern Territory, Australia (2013)

Factbook: Australia. Central Intelligence Agency, In: The world factbook. Available: https://www.cia.gov/library/publications/the-world-factbook/geos/AS.html (2016)

Georgees, R.N., et al.: Effect of the use of a polymeric stabilizing additive on unconfined compressive strength of soils. Transp. Res. Rec. J. Transp. Res. Board, 200–208 (2015)

Jones, D.: Development of performance-based tests for nontraditional road additives. Transp. Res. Rec. J. Transp. Res. Board, 142–153 (2007)

Malik, M., et al.: Mobility of polyacrylamide and polysaccharide polymer through soil materials. Soil Technol. **4**, 255–263 (1991)

Puppala, A., et al.: Permanent deformation characterization of subgrade soils from RLT test. J. Mater. Civ. Eng. **11**, 274–282 (1999)

Sampson, L.: Development of the mechanical wet/dry brushing test for the durability testing of stabilized materials. Transportek, CSIR, Pretoria, SA. Confidential Report C/DPVT/3.1 (1988)

Theng, B.: Clay-polymer interactions; summary and perspectives. Clays Clay Miner. **30**, 1–10 (1982)

Virgil Ping, W., et al.: Measuring resilient modulus of granular materials in flexible pavements. Transp. Res. Rec. J. Transp. Res. Board, 81–90 (2001)

Vorobieff, G., Wilmot, T.: Australian experience on subgrade stabilisation and pavement recycling. In: International Symposium on Subgrade Stabilisation and In Situ Pavement Recycling Using Cement, 1st, Salamanca, Spain (2001)

Wilmot, T.D.: Applications for stabilisation in pavement construction and rehabilitation. In: Malaysian Road Conference, 1st Kuala Lumpur, Malaysia (1994)

Analytical Method to Evaluate the Stress State Within Vertical Backfill

Qizhi Chen[1(✉)], Changjie Xu[1,2], Luju Liang[1], and Xiaozhen Fan[1]

[1] Research Center of Coastal and Urban Geotechnical Engineering, Zhejiang University, Hangzhou, China
chenqizhi@zju.edu.cn

[2] College of Civil Engineering and Architecture, East China Jiaotong University, Nanchang, China

Abstract. In mining industry, voids created during ore removal are generally backfilled with mill tailings that are left over from processing of mineral ore. Evaluating the stress state in backfill is a critical step for design of backfilled stopes. An analytical method to evaluate the stress is proposed in this paper. Nonuniform distribution of vertical stress is considered in this method. The trajectory of minor principle stress in the fill from the centreline to the wall is assumed as circular-arc. The stress state within vertical backfill can be obtained through the equilibrium consideration of differential flat element. Several examples carried out by analytical and numerical modeling are then presented and compared with existing analytical solutions. The proposed analytical method is also used to evaluate an in situ project from literature. The results by presented analytical method are in good agreement with numerical results and in situ data from literature. So the proposed analytical method can be used to evaluate the stress state in backfilled stopes.

1 Introduction

Mining is a multibillion dollar industry in countries such as Australia, United States, Canada, China, and India. In underground mining, the voids created during ore removal are called stopes and are generally backfilled with mill tailings that are left over from processing of mineral ore. To ensure safety within the access tunnels and other regions of the mines, it is necessary to determine the stress state within the backfilled stopes.

Arching always occurs when some load transfer along the interfaces induced by the deformation of backfilled materials. Friction shear stress acting on the interface of the wall and the granular material leads to arching, which leads in lower vertical stresses at any depths comparing to overburden stresses. A significant vertical stress reduction in a silo compared with the self-weight of grains was observed by Janssen (1895). Based on the limit equilibrium method, Janssen developed an expression to estimate the average vertical stress. Then in 1930, Janssen's approach was introduced to geotechnical engineering to evaluate the vertical stress in backfilled trenches by Marston.

Finite Differences Methods such as Fast Lagrangian Analysis of Continua (FLAC) are widely used to evaluate the stress state of granular material in a rigid container. Li et al. (2003) investigate the mechanical response of backfilled stopes with FLAC-2D.

© Springer International Publishing AG, part of Springer Nature 2019
A. Sevi et al. (eds.), *Enhancements in Applied Geomechanics, Mining, and Excavation Simulation and Analysis,* Sustainable Civil Infrastructures,
https://doi.org/10.1007/978-3-319-95645-9_8

Widisinghe and Sivakugan (2014, 2015) proposed the vertical stress isobars for trenches containing granular soils and the vertical stress isobars for silos containing food grains and square backfilled mine stopes containing mine fill. Other numerical modeling methods such as Finite Element Methods and Discrete Element Method are also used. Goodey et al. (2006) investigated state of stress in the stored solid and the pressures imposed on the silo walls using a Finite Element Model. Hasan et al. (2016) investigated the distribution of intergranular forces within uncemented mine backfills using the Discrete Element Method.

Analytical methods serve as basis for stress calculations in designs. Analytical methods also have the advantage of simplicity. However, analytical methods are always suffered from some limitations due to the simplifying assumptions. The numerical modelling results (Li et al. 2003) have shown that a uniform vertical stress across the width is not a valid hypothesis. For instance, most current existing solutions assume that both the horizontal and vertical stress across the width is uniform (Marston 1930; Krynine 1945; Pirapakaran and Sivakugan 2007; Ting et al. 2010).

This paper presents an analytical method to evaluate the stress state within vertical backfilled stopes. In this method, the nonuniform distribution of vertical stress across the width of backfill is considered. The trajectory of minor principle stress in the fill from the centreline to the wall is assumed as circular-arc. The direction of minor principle stress at centreline is vertical to the centreline, and the direction along the rock mass can be determined using the geometrical relationship in the Mohr's stress circle. The stress state of differential flat element in backfill is analyzed through equilibrium consideration. Then the analytical solution of stress state in backfill can be obtained. A series of examples calculated by presented analytical method and numerical modeling using ABAQUS are then presented, comparison between analytical results and in situ data is also presented and several conclusions can be drawn.

2 Analytical Method

The schematic representation of a vertical backfilled stope is shown in Fig. 1. The height and half width of the backfill are H and B, respectively. The backfilled soil is assumed as homogeneous, isotropic, cohesionless Mohr-Coulomb material. The unit weight is γ, and the internal friction angel is φ. Two vertical parallel walls of rock mass are assumed rigid and rough. The settlement of the backfilled soil is large enough to induce friction between the walls and the soil. The friction angle δ of the interface between the backfill and the rock mass is less than or equal to the backfill internal friction angle. A differential flat element with sickness dz at depth z is analyzed. As shown in Fig. 1, the major principle stress on a differential flat element are applied normal to the circular-arc (Paik and Salgado 2003) presented by the dotted lines, whereas the minor principal stresses are tangential to the direction of the circular-arc, becoming horizontal at the center of the differential element. Li et al. (2003) investigated the mechanical response of backfilled stopes with FLAC-2D simulations. Results show that horizontal normal stress component in backfill soils is same along width. The assumption that horizontal normal stress keeps constant along width is adopted in the following analytical solution.

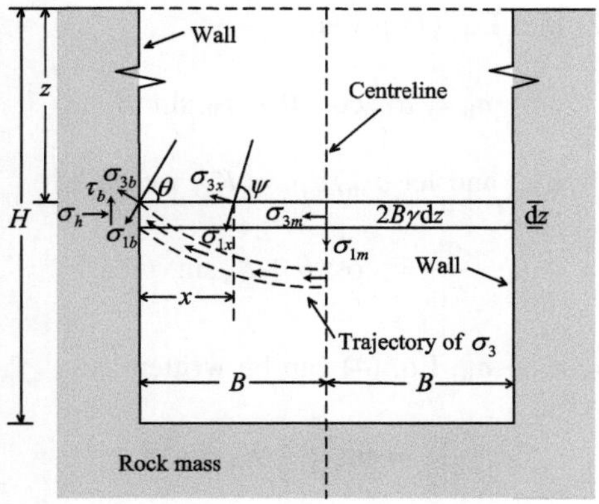

Fig. 1. Schematic representation of a vertical backfilled stope

The stress state of soils on the edge of differential flat element can be obtained according to the geometrical relationship in the Mohr's circle as shown in Fig. 2. The expressions of the shear stress and the horizontal stress:

$$\tau_b = (\sigma_{1b} - \sigma_{3b}) \sin\theta \cos\theta \tag{1}$$

$$\sigma_h = \tau_b \cot\theta + \sigma_{3b} \tag{2}$$

where $\theta = 45° + \varphi/2$, τ_b = shear stress on the edge of differential flat element, σ_{1b} = major principle stress on the edge of differential flat element, σ_{3b} = minor principle stress on the edge of differential flat element, σ_h = horizontal stress on differential flat element.

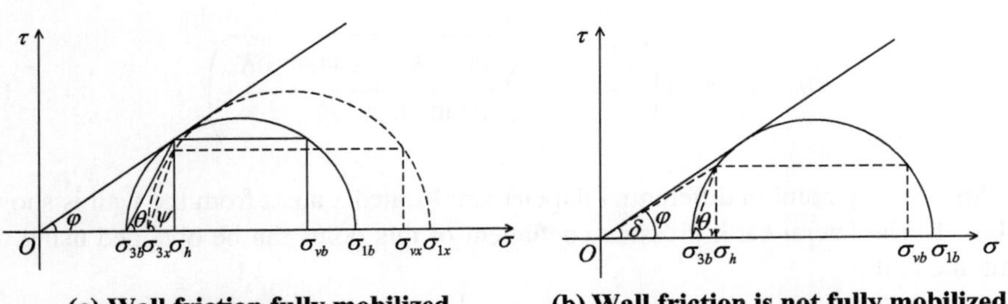

Fig. 2. Mohr's stress circle of cohesionless soils

Substituting Eq. (2) into Eq. (1) gives:

$$\sigma_h = \sigma_{1b}\cos^2\theta + \sigma_{3b}\sin^2\theta \tag{3}$$

Dividing Eq. (3) by σ_{1b}, and let $\sigma_{3b}/\sigma_{1b} = K_a$, we have:

$$\frac{\sigma_h}{\sigma_{1b}} = \cos^2\theta + K_a\sin^2\theta \tag{4}$$

Because $\sigma_1 + \sigma_3 = \sigma_{vb} + \sigma_h$, Eq. (4) can be written as:

$$\frac{\sigma_{vb}}{\sigma_{1b}} = \sin^2\theta + K_a\cos^2\theta \tag{5}$$

where σ_{vb} = vertical stress on the edge of differential flat element. Dividing Equation (4) by Equation (5) we can get the expression of the lateral earth pressure coefficient on the edge of differential flat element:

$$K_b = \frac{\sigma_h}{\sigma_{vb}} = \frac{\cos^2\theta + K_a\sin^2\theta}{\sin^2\theta + K_a\cos^2\theta} \tag{6}$$

The friction angle of the interface between the backfill and the rock mass is equal or less than the backfill internal friction angle. When the wall friction angle is less than the internal friction angle, as shown in Fig. 2b. The shear stress at the interface can be written:

$$\tau_b = (\sigma_h + \sigma_{3b})\tan\theta_w = \sigma_h\tan\delta \tag{7}$$

The angle between minor principle stress and vertical direction at the interface can be obtained as:

$$\theta_w = \arctan\left(\frac{1 - K_a + \sqrt{(1-K_a)^2 - 4\tan^2\delta K_a}}{2\tan\delta K_a}\right) \tag{8}$$

An arbitrary point in differential flat element located x apart from the wall is shown in Fig. 1. The lateral earth pressure coefficient of this point can be obtained using the same method:

$$K_x = \frac{\sigma_h}{\sigma_{vx}} = \frac{\cos^2\psi + K_a\sin^2\psi}{\sin^2\psi + K_a\cos^2\psi} \tag{9}$$

where ψ = angle between minor principle stress and vertical direction.

The vertical static equilibrium equation of the differential flat element in Fig. 1 can be written as:

$$2B\gamma \mathrm{d}z = 2B\mathrm{d}\overline{\sigma_{\mathrm{v}}} + 2K_{\mathrm{b}}\sigma_{\mathrm{vb}} \tan\varphi \mathrm{d}z \tag{10}$$

The angle φ in Eq. (10) can be substituted by δ, if friction between the walls and the soil is not fully mobilized.

Average vertical stresses acting on the differential flat element can be written as:

$$\overline{\sigma_{\mathrm{v}}} = \frac{\int_0^{2B} \sigma_{\mathrm{vx}} \mathrm{d}x}{2B} = \frac{\sigma_{\mathrm{vb}} \cdot K_{\mathrm{b}}}{B} \int_0^{B} \frac{1-(1-K_{\mathrm{a}})\cos^2\psi}{(1-K_{\mathrm{a}})\cos^2\psi + K_{\mathrm{a}}} \mathrm{d}x \tag{11}$$

The relationship between ψ and θ_{w} for circular-arc trajectory of minor principle stress is shown as:

$$\cos\psi = \frac{B-x}{B}\cos\theta_{\mathrm{w}} \tag{12}$$

Letting m be the shape coefficient of minor principal stress trajectory, we have:

$$\overline{\sigma_{\mathrm{v}}} = m\sigma_{\mathrm{vb}} \tag{13}$$

Substituting Eqs. (12), (13), and (14) into Eq. (11) respectively, the shape coefficients of circular-arc are obtained, as shown:

$$m_{\mathrm{cir}} = K_b \cdot \left[\frac{1 - K_{\mathrm{a}} \cdot \arctan\left(\frac{\cos\theta_{\mathrm{w}} \cdot \sqrt{1-K_{\mathrm{a}}}}{\sqrt{K_{\mathrm{a}}}}\right)}{\cos\theta_{\mathrm{w}} \cdot \sqrt{K_{\mathrm{a}}(1-K_{\mathrm{a}})}} - 1 \right] \tag{14}$$

Substituting Eq. (13) into Eq. (10) gives:

$$2Bm\gamma \mathrm{d}z = 2Bm\mathrm{d}\overline{\sigma_{\mathrm{v}}} + 2K_{\mathrm{b}}\overline{\sigma_{\mathrm{v}}} \tan\varphi \mathrm{d}z \tag{15}$$

Using the boundary condition on the top of backfill, we have:

$$z = 0, \overline{\sigma_{\mathrm{v}}} = 0 \tag{16}$$

Equation (13) can be solved as:

$$\overline{\sigma_{\mathrm{v}}} = \frac{\gamma Bm}{K_{\mathrm{b}} \tan\varphi}\left(1 - \mathrm{e}^{-\frac{zK_{\mathrm{b}}\tan\varphi}{mB}}\right) + q\mathrm{e}^{-\frac{zK_{\mathrm{b}}\tan\varphi}{mB}} \tag{17}$$

Then with Eqs. (13) and (17), the expression of vertical stress of an arbitrary point in backfill can be derived as:

$$\sigma_{\mathrm{vx}} = \frac{K_{\mathrm{b}}}{K_{\mathrm{x}}} \cdot \frac{\overline{\sigma_{\mathrm{v}}}}{m} \tag{18}$$

3 Analytical and Numerical Examples

Analytical and numerical examples are presented in this section. Numerical examples are done by Finite Element software ABAQUS. Mohr-Coulomb plasticity model is used to model the backfill soils and rigid body is used to model the rock mass. Young's modulus of the backfill soils is 300 Mpa and Poisson's ratio is 0.2. The unit weight of backfill is 18 kN/m^3. From geotechnical consideration the angle of wall friction is slightly less than backfill internal friction angle. Pirapakaran and Sivakugan (2007) used the value that $\delta = 2\varphi/3$ to estimate the vertical stress in a trench. In the proposed method, the value that $\delta = 0.9\varphi$ is used.

Several analytical and numerical results are shown in Fig. 3 for $H = 32$ m, $B = 6$ m, and $\varphi = 30°$. The results from previous analytical solutions in literature (Marston 1930; Krynine 1945) are also shown in Fig. 3. It can be seen from Fig. 3a–d that the results done by proposed method fit well with numerical results. For stress along the centerline, results by proposed method, Marston's method and numerical method have a good agreement for both vertical and horizontal stress. Krynine's method failed to estimate the stress state along the centerline. For stress state along the wall, results by proposed method, Krynine's method and numerical method have a good agreement for vertical stress, and results by proposed method, Marston's method and numerical method have a good agreement for horizontal stress. However Marston's method failed to calculate the vertical stress along the wall. Figure 3e and g shows a nonuniform distribution of vertical stress which has not been considered in previous studies. The vertical stress near the wall is much smaller than that near the centerline. From Fig. 3f and h we can see that the horizontal stress by numerical results across the width nearly keeps constant, which is accord with the assumption in proposed method.

Other analytical and numerical examples for different width of backfill and different internal friction of backfilled materials are also done. However the results are not shown in this paper because of the limitation of paper length.

Knutsson (1981) carried out in situ stress measurements within backfill stopes at Nasliden mine of the Boliden Company for several years. The in situ data from Knutsson is also compared with the proposed analytical method for verification purposes. Knutsson's in situ data is reproduced after Pirapakaran (2008), which contains vertical and horizontal stresses along the depth of the backfill. The measured stope is 6 m wide, 45 m high, and 160 m long, which can be considered as plane strain condition. The unit weight of the in situ backfill is 21.6 kN/m^3. The internal friction angle is 30°. The comparisons between results by proposed analytical method and in situ data are shown in Fig. 4. A good agreement can be obtained between measured data and results by proposed method. At the stope bottom, calculated stresses are obviously less than overburden pressure due to arching effect, which is verified by measured data.

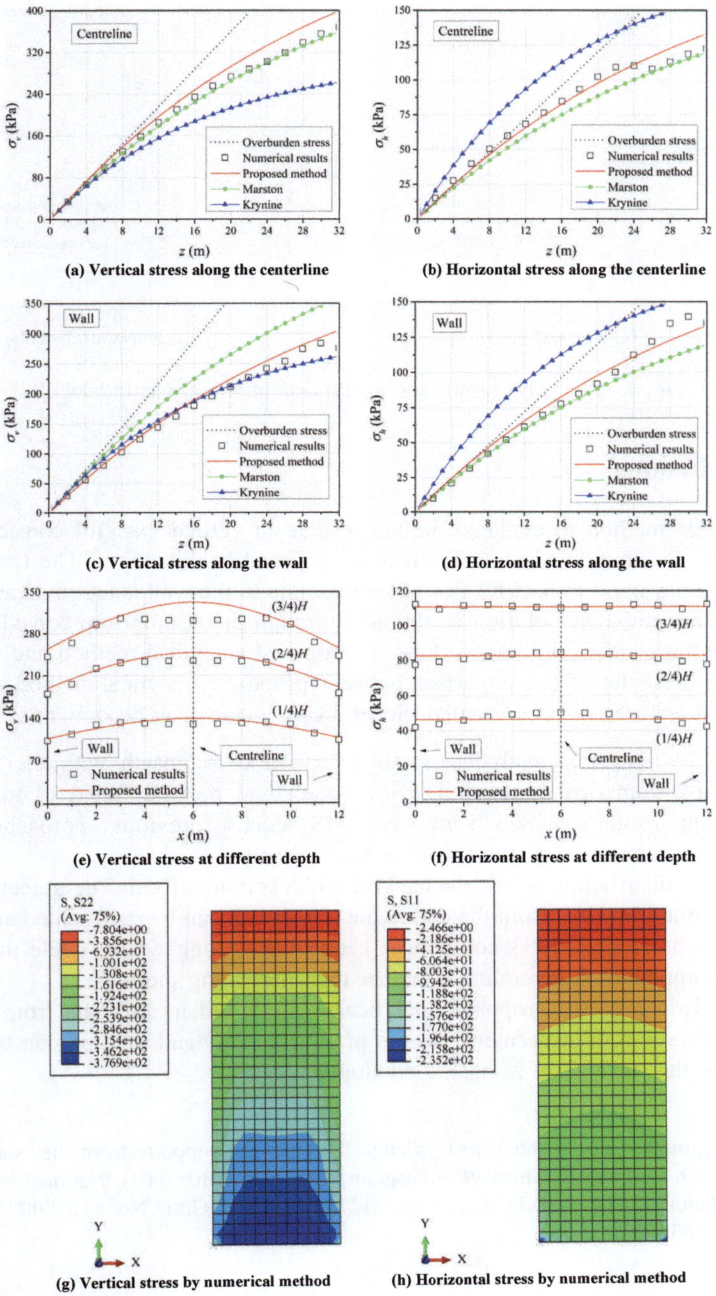

Fig. 3. Analytical and numerical results of stress state in backfill

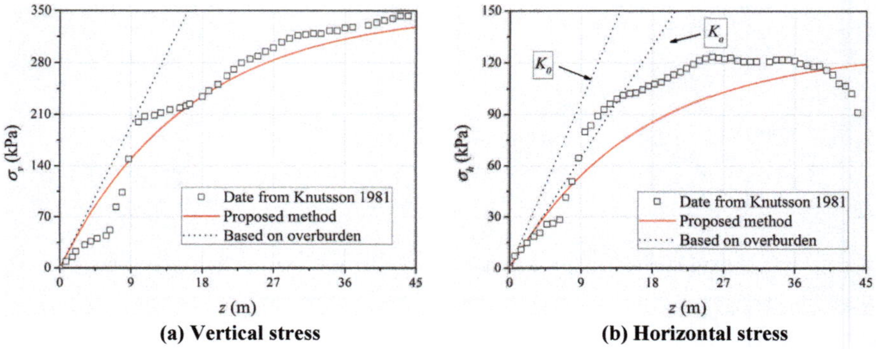

Fig. 4. Analytical results and in situ data of stress state in backfill

4 Conclusions

An analytical method to evaluate the stress state in vertical backfill considering the nonuniform distribution of vertical stress is proposed in this paper. The trajectory of minor principle stress in backfill from the centreline to the wall is assumed as circular-arc. Then the analytical solution is obtained through the equilibrium consideration of differential flat element. Examples done by proposed analytical method and numerical method are presented. The comparison between proposed analytical method and in situ data from literature is also presented. Several conclusions can be drawn:

1. Results by proposed analytical method have a close match with the numerical results by numerical method. Previous analytical method failed to match with numerical results obtained from ABAQUS because previous approaches cannot present the nonuniform stress state of backfill soils.
2. The stress distribution across the backfill width is nonuniform. The trajectory of the minor principle stress from the centreline to the wall can be represented as circular-arc. The horizontal stress component keeps same along width, while the vertical stress component at centreline is higher than that along the wall.
3. Comparison between proposed analytical method and in situ data from literature also shows a good agreement. So the proposed analytical method can be used to evaluate the stress state in backfilled stopes.

Acknowledgments. The authors acknowledge the financial supports from the National Key Basic Research Program of China (973 Program) (No. 2015CB057801), National Natural Science Foundation of China (NSFC Grant No. 51278449, NSFC Grant No. 51238009 and NSFC Grant No. 51338009).

References

ABAQUS User's Manual, Version 6.1, Hibbitt, Karlsson and Sorenson Inc. (2000)

Goodey, R.J., Brown, C.J., Rotter, J.M.: Predicted patterns of filling pressures in thin-walled square silos. Eng. Struct. **28**(1), 109–119 (2006)

Hasan, A., Karrech, A., Chareyre, B.: Evaluating force distributions within virtual uncemented mine backfill using discrete element method. Int. J. Geomech. **17**(7), 06016042 (2016)

Janssen, H.A.: Versuche über getreidedruck in silozellen. Zeitschr. d. Vereines deutscher Ingenieure **39**(35), 1045–1049 (1895)

Knutsson, S.: Stresses in the hydraulic backfill from analytical calculations and in-situ measurements. In: Conference on the Application of Rock Mechanics to Cut and Fill Mining: 01/06/1980-03/06/1980, pp. 261–268. The Institution of Mining and Metallurgy (1981)

Krynine, D.P.: Discussion on 'Stability and stiffness of cellular cofferdams' by Karl Terzaghi. Trans. Am. Soc. Civ. Eng. **110**(1), 1175–1178 (1945)

Li, L., Aubertin, M., Simon, R., Bussière, B., Belem, T.: Modeling arching effects in narrow backfilled stopes with FLAC. In: Proceedings of the 3rd International Symposium on FLAC & FLAC 3D Numerical Modelling in Geomechanics, Ontario, Canada, pp. 211–219 (2003)

Marston, A.: The theory of external loads on closed conduits in the light of the latest experiments. In: Highway Research Board Proceedings, vol. 9. (1930)

Paik, K.H., Salgado, R.: Estimation of active earth pressure against rigid retaining walls considering arching effects. Géotechnique **53**(7), 643–653 (2003)

Pirapakaran, K., Sivakugan, N.: Arching within hydraulic fill stopes. Geotech. Geol. Eng. **25**(1), 25–35 (2007)

Pirapakaran, K.: Load-deformation characteristics of minefills with particular reference to arching and stress developments. Doctoral dissertation, James Cook University (2008)

Ting, C.H., Sivakugan, N., Read, W., Shukla, S.K.: Analytical method to determine vertical stresses within a granular material contained in right vertical prisms. Int. J. Geomech. **12**(1), 74–79 (2010)

Widisinghe, S., Sivakugan, N.: Vertical stress isobars for trenches and mine stopes containing granular backfills. Int. J. Geomech. **14**(2), 313–318 (2014)

Widisinghe, S., Sivakugan, N.: Vertical stress isobars for silos and square backfilled mine stopes. Int. J. Geomech. **16**(2), 06015003 (2015)

Numerical Simulation of Surface Subsidence After the Collapse of a Mine

Y. G. Derbin[1,2(✉)], J. Walker[1,2], D. Wanatowski[2,3], and A. M. Marshall[2,4]

[1] University of Nottingham, Ningbo, China
Yury.Derbin@nottingham.edu.cn
[2] State Key Laboratory for GeoMechanics and Deep Underground Engineering, China University of Mining and Technology, Xuzhou, China
[3] University of Leeds, Leeds, UK
[4] University of Nottingham, Nottingham, UK

Abstract. Surface subsidence is a concern for many underground mining activities. If not predicted, this phenomenon can cause severe infrastructure damage. In this paper, a computer model is used to predict surface subsidence after the controlled collapse of a coal mine at Naburn in North Yorkshire, England. Scarcity of data on the characteristics of deep underground distressed and caved zones around coal mining excavations makes the numerical prediction of mining-induced subsidence very difficult. The authors derive appropriate input parameters for the numerical model using available borehole data with all necessary justifications provided. Simulations are performed using the commercial software FLAC3D. Different constitutive models, such as Mohr-Coulomb, modified Hoek-Brown, strain-softening, double-yield, and modified Cam-clay are used to obtain surface subsidence profiles, which are compared against measurements taken at the site. Special attention is given to numerically simulating processes involved in the underground movements. It is shown that none of the models listed above can reasonably predict the surface subsidence profile.

Keywords: Coal mining · Numerical modelling · Surface subsidence
Caved zone · Distressed zone

1 Introduction

Any underground work can cause surface subsidence, which could damage infrastructure and buildings. Longwall coal mining is not an exception. In order to mitigate consequences of the subsidence and choose the appropriate method of mining, it is important to predict the size and depth of the trough. Some empirical methods of surface subsidence prediction have been developed in different countries based on the observed local data; for example, one of them is in the Subsidence Engineers' Handbook (NCB 1975) developed in the UK. FLAC3D is commercial software, which has been used to predict the surface subsidence trough (Herrero et al. 2012 and Xu et al. 2013). However, due to the complex behaviour of the rock, there is minimal confidence

© Springer International Publishing AG, part of Springer Nature 2019
A. Sevi et al. (eds.), *Enhancements in Applied Geomechanics, Mining, and Excavation Simulation and Analysis*, Sustainable Civil Infrastructures,
https://doi.org/10.1007/978-3-319-95645-9_9

in predictions from numerical modelling and more research is required (Xu et al. 2013). The purpose of this work is to increase the understanding and develop the procedure of the numerical simulation of the surface subsidence with limited information on the properties of the overburden.

The following procedure are followed:

- Deriving model parameters from the borehole data;
- Setting the domain of the model;
- Assigning the constitutive model;
- Assigning the parameters derived earlier;
- Setting initial and boundary conditions;
- Running the model to equilibrium;
- Assigning the special constitutive model to the excavating zone to model goaf behavior;
- Assigning the parameters, which are inherited from the roof layer or derived from the literature, to the goaf;
- Assigning characteristics of the volumetric behavior of the goaf;
- Running the model to equilibrium;
- Altering volumetric characteristics in the goaf to obtain the required goaf height after the simulation (adjusting the goaf height).

The paper explains the procedure and discusses the modelling results in three parts. The first part explains a method developed by the authors for estimating and assigning the appropriate physical-mechanical properties to the model using a visual description of the borehole log. This method was created after an extensive literature search and based on the works of Balmer (1952), Deere (1968), Hoek and Brown (1980), Hansen (1988), Palmström (1995), Hoek and Brown (1997), Palmström and Singh (2001), and Hoek and Diederichs (2006). The second part describes the site of interest, the model domain, the mesh density, the initial and boundary conditions. The site is at Naburn in North Yorkshire, England. A measured subsidence profile was obtained from UKCoal. The mine collapsed controllably after Longwall mining without stowing. The third part discusses results, the modelled stresses in the goaf, the debris caused by a mine collapse, the influence of the Geological Strength Index (GSI) and surface subsidence.

2 Physical-Mechanical Properties

2.1 Elastic Stiffness

Before developing a surface subsidence model, the estimation of the strength and deformation characteristics of rock masses should be carried out. It can be done in the laboratory, but it is expensive and suitable samples are not always available. After a wide literature review, a method for the estimation of the rock properties based on the borehole log description was developed and is presented in this section.

A key property of the material is the elastic stiffness. Hoek and Diederichs (2006) estimated it by the formula

$$E = E_i\left(0.02 + \frac{1 - D/2}{1 + e^{(60 + 25D - GSI)/11}}\right) \quad (1)$$

where GSI is the Geological Strength Index introduced by Hoek and Brown (1997) and describes the insitu state of the rock. A GSI of 100 is for a very good, undisturbed rock mass whereas a GSI of 0 is for a very poor quality, disintegrated rock mass. D is the disturbance factor, which is dependent upon the excavation conditions, i.e. blasting a rock face will give the rock a disturbance factor of 1 while careful excavation will yield a disturbance factor of 0. E_i is given by:

$$E_i = MR \cdot \sigma_{ci} \quad (2)$$

where MR is the modulus ratio classified by rock type and presented in Table 1, MR was first proposed by Deere (1968) and later modified by Palmström and Singh (2001). σ_{ci} is the uniaxial compressive strength of the intact rock, which could be found in Table 2.

Table 1. Guidelines for the selection of the modulus ratio. Based on Deere (1968) and Palmström and Singh (2001)

Rock type	Class	Group	Texture			
			Coarse	Medium	Fine	Very fine
Sedimentary	Clastic		Conglome-rates 300–400 Breccias 230–350	Sandstones 200–350	Siltstones 350–400 Greywackes 350	Claystones 200–300 Shales 150–250[a] Marls 150–200
	Non-clastic	Carbonates	Crystalline limestones 400–600	Sparitic limestones 600–800	Micritic limestones 800-1000	Dolomites 350–500
		Evaporites		Gypsum (350)[b]	Anhydrite (350)[b]	
		Organic				Chalk 1000+
Metamorphic	Non-foliated		Marble 700–1000	Hornfels 400–700 Metasandstone 200–300	Quartzites 300–450	
	Slightly foliated		Migmatite 350–400	Amphibolites 400–500	Gneiss 300–750[a]	
	Foliated[a]			Schists 250–1100[a]	Phyllites/mica schist 300–800[a]	Slates 400–600[a]
Igneous	Plutonic	Light	Granite[c] 300–550	Diorite[c] 300–350		
			Granodiorite[c] 400–450			
		Dark	Gabbro 400–500	Dolerite 300–400		
			Norite 350–400			
	Hypabyssal		Porphyrics (400)[b]		Diabase 300–350	Peridotite 250–300
	Volcanic	Lava		Rhyolite 300–500 Andesite 300–500	Dacite 350–450 Basalt 250–450	
		Pyroclastic	Agglomerate 400–600	Volcanic breccia (500)[b]	Tuff 200–400	

[a]Highly anisotropic rocks: the value of MR will be significantly different if normal strain and/or loading occurs parallel (high MR) or perpendicular (low MR) to a weakness plane. Uniaxial test loading direction should be equivalent to field application
[b]No data available, estimated on the basis of geological logic
[c]Felsic Granitoids: coarse grained or altered (high MR), fined grained (low MR)

Table 2. A summary of uniaxial compressive strength based on rock type (Palmström 1995)

Rock name	Uniaxial compressive strength σc, MPa			Rating of the factor m_i[1)]	Rock name	Uniaxial compressive strength σc, MPa			Rating of the factor m_i[1)]
	Low	Average	High			Low	Average	High	
Sedimentary rocks					*Metamorphic rocks*				
Anhydrite	120[b]?			13.2	Amphibolite	75	125	250	31.2
Coal	16[c]	21[c]	26[c]		Amphibolitic gneiss	95	160	230	31?
Claystone	2[b]	5[b]	10[b]	3.4	Augen gneiss	95	160	230	30?
Conglomerate	70	85	100	(20)	Black shale	35	70	105	
Coral chalk	3	10	18	7.2	Garnet mica schist	75	105	130	
Dolomite	60[b]	100[b]	300[b]	10.1	Granite gneiss	80	120	155	30?
Limestone	50[a]	100[b]	180[a]	8.4	Granulite	80[b]	150	280	
Mudstone	45	95	145		Gneiss	80	130	185	29.2
Shale	36[c]	95[c]	172[c]		Gneiss granite	65	105	140	30?
Sandstone	75	120	160	18.8	Greenschist	65	75	85	
Siltstone	10[b]	80[b]	180[b]	9.6	Greenstone	120[b]	170[a]	280[a]	20?
Tuff	3[b]	25[b]	150[b]		Greywacke	100	120	145	
Igneous rocks									
Andesite	75[b]	140[b]	300[b]	18.9	Marble	60[b]	130[b]	230[b]	9.3
Anorthosite	40	125	210		Mica gneiss	55	80	100	30?
Basalt	100	165	355[c]	(17)	Mica quartzite	45	85	125	25?
Diabase (dolerite)	227[c]	280[c]	319[c]	15.2	Mica schist	20	80[a]	170[a]	15?
Diorite	100	140	190	27?	Mylonite	65	90	120	
Gabbro	190	240	285	25.8	Phyllite	21	50	80	13?
Granite	95	160	230	32.7	Quartz sandstone	70	120	175	
Granodiorite	75	105	135	20?	Quartzite	75	145	245	23.7
Monzonite	85	145	230	30?	Quartzitic phyllite	45	100	155	
Nepheline syenite	125	165	200		Serpentinite	65	135	200	
Norite	290[c]	298[c]	326[c]	21.7	Slate	120[b]	190[b]	300[b]	11.4
Pegmatite	39	50	62						
Rhyolite		85[b]?		(20)	Talc schist	45	65	90	10?
Syenite	75	150	230	30?					
Ultra basic rock	80[b]	160	360						

Soil materials[2)]

Very soft clay σ_c = 0.025 MPa Soft clay σ_c = 0.025 − 0.05 MPa Firm clay σ_c = 0.05 − 0.1 MPa

Stiff clay σ_c = 0.1 − 0.25 MPa Very stiff clay σ_c = 0.25 − 0.5 MPa Hard clay σ_c => 0.5 MPa

Silt, sand: assume σ_c = 0.0001 − 0.001 MPa

[a]Values found by the Technical University of Norway, (NTH) Inst. for rock mechanics

[b]Values given in Lama and Vutukuri, 1978

[c]Values given by Bieniawski, 1984

[1)]mi refers to the value of m for intact rock in the Hoek-Brown model. Values in brackets have been estimated by Hoek et al. (1992) while those with a question mark have been assumed by Palmstrom

[2)]For clays, the values of the UCS is based on ISRM (1988)

2.2 Failure Parameters

Hansen (1988) and Hoek and Brown (1980) developed the algorithm to determine failure parameters based upon a description of a borehole log. In order to find the Mohr-Coulomb parameters, cohesion and internal friction, the Hoek-Brown failure criterion is approximated with a Mohr-Coulomb failure surface by following the work of Hoek and Brown (1997). Equation 3, the Hoek-Brown empirical failure criterion for jointed rock masses, provides values of σ_1 that lay on the yield surface for different values of σ_3 hence it can be used to generate the maximum and minimum principal stresses for the rock in question.

$$\sigma_1^{'} = \sigma_3^{'} + \sigma_{ci}\left(m_b \frac{\sigma_3^{'}}{\sigma_{ci}} + s\right)^{0.5} \tag{3}$$

Correspondingly, the values of m_b, and s in Eq. 3 are given by:

$$m_b = m_i.\exp\left(\frac{GSI - 100}{28}\right) \quad \text{if GSI} > 25 \tag{4}$$

where, m_i refers to the value of m for intact rock in the Hoek-Brown model and is summarized in Table 3, and

$$s = \exp\left(\frac{GSI - 100}{9}\right) \tag{5}$$

Table 3. Determination of m_i (Hoek and Brown 1997)

Rock type	Class	Group	Texture			
			Coarse	Medium	Fine	Very fine
Sedimentary	Clastic		Conglomerate (22)	Sandstone 19	Siltstone 9	Claystone 4
				Greywacke (18)		
	Non-clastic	Organic		Chalk 7 Coal (8–21)		
		Carbonate	Breccia (20)	Sparitic limestone (10)	Micritic limestone 8	
		Chemical		Gypstone 16	Anhydrite 13	
Metamorphic	Non-foliated		Marble 9	Hornfels (19)	Quartzite 24	
	Slightly foliated		Migmatite (30)	Amphibolite 25–31	Mylonites (6)	
	Foliated[a]		Gneiss 33	Schists 4–8	Phyllites (10)	Slate 9
Igneous	Light		Granite 33 Granodiorite (30) Diorite (28)		Rhyolite (16) Dacite (17) Andesite 19	Obsidian (19)
	Dark		Gabbro 27 Norite 22	Dolerite (19)	Basalt (17)	
	Extrusive pyroclastic type		Agglomerate (20)	Breccia (18)	Tuff (15)	

[a]These values are for intact rock specimens tested normal to bedding or foliation. The value m_i will be significantly different if failure occurs along a weakness plane

Then, the values for σ_3 must be selected. Hoek and Brown (1997) concluded that the most consistent results are obtained when 8 equally spaced values between $0 < \sigma_3 < 0.25\sigma_{ci}$ are used.

To find the tangent of the failure surface at the appropriate stress level, first the non-linear analytical solution for Mohr's envelope is found and then a linear regression analysis is used to find the equation of the tangent at that point.

Balmer's analytical solution (Balmer 1952) to Mohr's envelope describes the relationship between the normal and shear stresses in terms of the principal stresses as:

$$\sigma_n' = \sigma_3' + \frac{\sigma_1' - \sigma_3'}{\partial\sigma_1'/\partial\sigma_3' + 1} \tag{6}$$

$$\tau = \left(\sigma_1' - \sigma_3'\right)\sqrt{\partial\sigma_1'/\partial\sigma_3'} \tag{7}$$

Provided that the GSI is greater than 25 we can calculate

$$\frac{\partial\sigma_1'}{\partial\sigma_3'} = 1 + \frac{m_b\sigma_{ci}}{2(\sigma_1' - \sigma_3')} \tag{8}$$

The tensile strength of the rock is calculated by substitution of $\sigma_1 = 0$ and $\sigma_{tm} = -\sigma_3$ in the Hoek-Brown failure criterion (Eq. 3):

$$\sigma_{tm} = \frac{\sigma_{ci}}{2}\left(m_b - \sqrt{m_b^2 + 4s}\right) \tag{9}$$

The equivalent Mohr envelope may be written as:

$$Y = \log(A) + B \cdot X \tag{10}$$

which requires determination of A and B. The values of X and Y can be calculated using Eqs. 7, 8, and:

$$Y = \log\left(\frac{\tau}{\sigma_{ci}}\right) \tag{11}$$

$$X = \log\left(\frac{\sigma_n' - \sigma_{tm}}{\sigma_{ci}}\right) \tag{12}$$

The constants A and B can then be calculated using a linear regression analysis, i.e.

$$A = 10^{\left(\sum Y/T - B\left(\sum X/T\right)\right)} \tag{13}$$

$$B = \frac{\sum X \cdot Y - (\sum X \cdot \sum Y)/T}{(\sum X^2 - (\sum X^2)/T)} \tag{14}$$

where T is the number of values in the sequence, i.e. 8, if the earlier suggestion is followed.

Finally, the Mohr-Coulomb parameters can be deduced from the following two equations:

$$\phi = \mathrm{Tan}^{-1}\left(A \cdot B \cdot \left(\frac{\sigma'_{ni} - \sigma_{tm}}{\sigma_{ci}}\right)^{B-1}\right) \tag{15}$$

$$c = 0.75\left(\tau - \sigma'_{ni}\mathrm{Tan}(\phi)\right) \tag{16}$$

Since A and B are known, then using σ_{ni} in place of σ_n, (i.e. general notation of the normal stress), in Eq. 12 will evaluate an expression for X which can be used in Eq. 10 to find a value of Y. This can be used to calculate τ from Eq. 11.

It can be seen that another new parameter (σ_{ni}) has been introduced and this is the value of the normal stress at the point of interest. To determine this value we need to turn to the work of Hoek and Brown (1980).

Hoek and Brown found the correlation (Eq. 17) between depth and vertical insitu stress based on the collated worldwide data from researchers investigating the insitu state of stress underground (Fig. 1).

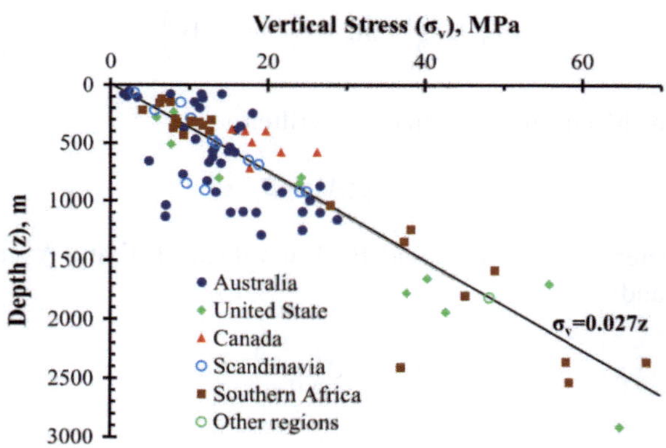

Fig. 1. Insitu vertical stress data. after Hoek and Brown (1980) in Hudson and Harrison (2000)

$$\sigma_{ni} = 0.027 \cdot z \tag{17}$$

The vertical stress calculated by Eq. 17 is given in MPa. This value is equivalent to the normal stress and enables us to completely specify the mechanical properties of the rocks underground.

3 Model Development

3.1 Site of Interest

A measured subsidence profile and borehole log description were taken from both above and under the Barnsley seam at Naburn in North Yorkshire, UK. The overburden consists of siltstone, sandstone, mudstone, and seatearth (claystone underlying coal seam). The bulk and shear moduli were calculated using Eqs. 1 and 2 for the elastic stiffness and assuming a Poisson's ratio of 0.2 for all layers. The internal friction and cohesion were derived from Eqs. 15 and 16. The tensile strength was calculated by Eq. 9. A density of 2700 kg/m^3 was taken as an average value for these types of rock after data collected by Shtumpf (1994) in Table 4.

Table 4. Density of different geomaterial (Shtumpf 1994)

Geomaterial type	Density (kg/cm^3)
Soils	2000–2780
Sandstone	2400–2900
Siltstone	2390–2950
Mudstone	2400–2800
Distorted and coal included rock	1800–2450
Sandstone and siltstone with siderite, pyrite, chalcopyrite and with other heavy metals	2950–3600

3.2 Model Domain

Due to the symmetry of the problem, the model domain could be reduced by half of the profile to reduce running time of the simulation. The size of domain was chosen in such a way that the boundary conditions did not impact on the result. Figure 2 shows the sizes of the model, the placement of the goaf, the location of the roller boundary conditions, and two different densities of the mesh. The excavation under investigation was 2.8 m thick, 75 m wide, and 709.6 m below the surface. The model was fixed in the out of plane direction. The bottom of the model was fixed in the vertical direction, and two sides were fixed in the horizontal direction. The lowest density mesh was located where the stresses were low following the recommendations of the FLAC3D manual (Itasca 2013). Either the Mohr-Coulomb, modified Hoek-Brown, or the strain-hardening/softening models constituted the behaviour of the whole subsurface. The exception was the goaf material, where the double-yield and later the modified Cam-clay models were implemented.

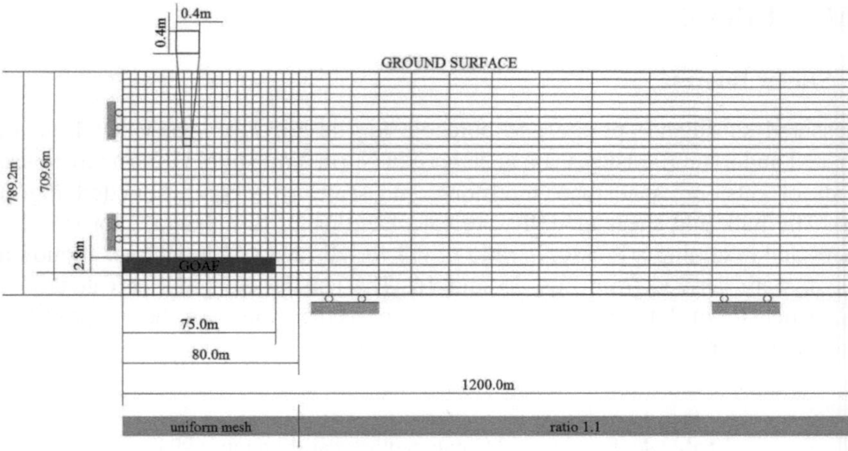

Fig. 2. Scheme of the model

3.3 Simulation of the Goaf Behaviour

According to Najafi et al. (2014), the simulation of the goaf behaviour is important for the accuracy of the subsidence prediction. Herewith, the goaf material is deep underground and it is difficult to estimate its properties. There have been numerous attempts at characterizing the goaf behaviour in the literature. For example, Salamon (1983) described the volumetric compression properties of the goaf material by the following equation:

$$\sigma = \frac{\alpha\varepsilon}{\beta - \varepsilon} \tag{18}$$

where α and β are empirical constants.

Later, Salamon (1990) rewrote Eq. 18 and eliminated the empirical constants by using certain physical parameters.

$$\sigma = \frac{E_0\varepsilon}{(1 - \varepsilon/\varepsilon_m)} \tag{19}$$

where E_0 is the initial tangent modulus and ε_m is the maximum strain of the goaf material.

Since the parameters are difficult to estimate, and even sometimes impossible, the authors of this paper go further with assumptions. Equation 18 can be rewritten considering the coefficient $\gamma = 1/\varepsilon_m$, which is used to adjust the height of the goaf after a simulation, and E is the Young's modulus of the roof material.

$$\sigma = \frac{E\varepsilon}{(1 - \gamma\varepsilon)} \tag{20}$$

The correctness of the modelled goaf behaviour is possible to check by two facts: the goaf height and the stresses in the goaf after the simulation. The stresses in the goaf will be discussed in the next section.

The required goaf height at the end of the simulation can be estimated by the initial height of the seam. The required height can be found by the multiplication between the height of the mine and the subsidence factor: these are collected in Table 5 for different regions of the world by Bräuner (in Bell and Donnelly 2006). The subsidence factor depends on the region and whether or not the excavated area has been filled or packed. The factor is used to calculate maximum possible subsidence. Table 5 shows the subsidence factor varies from 0.33 to 0.9 for the different regions. The recommended factor is 0.9 for mines in UK. For the case at hand with a height of the excavation of 2.8 m, the required final height of the goaf is estimated as 0.28 m. Altering the parameter γ, the simulation was repeated until the goaf height after the simulation becomes within 5% error of the required height. For the sake of simplicity, this process of obtaining the required goaf height by altering the volumetric characteristics of the goaf is called 'adjusting goaf height'.

Table 5. Subsidence factor after Bräuner (1973)

Coal field	Subsidence factor
British coal fields	0.90
Ruhr coal field, Germany	0.90
North and Pas de Calais coal field, France	0.85–0.90
Upper Silesia, Poland	0.70
Donbass district, Ukraine	0.80
Lvov-Volyn district, Ukraine	0.80–0.90
Kizelov district, Ukraine	0.40–0.80
Donetz, Kuznetsk and Karaganda districts, Russia and Ukraine	0.75–0.85
Sub-Moscow and Cheliabinski districts, Russia	0.85–0.90
Pechora, Russia	0.65–0.90
Central, USA	0.50–0.60
Western, USA	0.33–0.65

In order to model the strain-hardening behaviour of the goaf material, the double-yield model, which allows both shear and volumetric compression, is traditionally implemented. In FLAC, the stress-strain curve is approximated by a table to generate a linear piecewise curve. In the developed model, the table has 10 rows. The elastic properties, bulk and shear moduli, and Mohr-Coulomb properties, friction and cohesion, correspond to the properties of the roof material.

As it will be shown later, the double-yield model cannot simulate the goaf behaviour precisely enough. Instead, one of the Critical State models, namely the modified Cam-clay model, was implemented. Derbin et al. (2016) showed that the modified Cam-clay model predicted the goaf behaviour more accurately than the double-yield model did. Computational application of Critical State theory includes different types of soil and soft rock (Gens, and Potts 1988). Xiao et al. (2016) successfully implemented the Critical State concept to predict the behaviour of coarse granular soil (which is a material suitable for a rockfill dam) in a true triaxial compression test. The rockfill could best be described as a very coarse granular type of soil according to the British Soil Classification System (BS 5930: 1981). The very coarse soils are cobbles with sizes of 63–200 mm, boulders with sizes of 200–630 mm, and large boulders with sizes of more the 630 mm. Singh and Singh (2011) argued that goaf consists of 22.5% boulders and 77.5% large boulders. It should be also noted that the elastic properties depend on strain in the modified Cam-clay model. This corresponds better to the real behaviour of the goaf (Badr et al. 2003).

The goaf material was described by critical state parameters, i.e. lambda (λ) = 0.188, kappa (κ) = 0.007 and a frictional constant (M) = 1.9, which were used by Indraratna and Salim (2002) to model drain triaxial shearing on crushed basalt. By changing either the specific volume at reference pressure on the normal consolidation line or the pre-consolidation pressure, the required height of the goaf can be obtained. Figure 3 shows curves of dependence between the obtained goaf height and specific volume for three different pre-consolidation pressures, i.e. 1e5, 1e4, and 1e3 Pa.

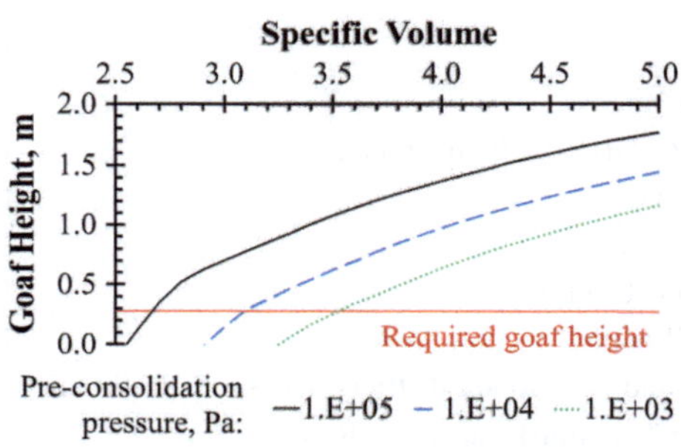

Fig. 3. Obtaining the required height of the goaf

4 Model Results

4.1 Surface Subsidence

The existing constitutive models that are available in almost all commercial software aren't capable of providing accurate solutions. Figure 4 depicts the surface settlement half-profiles. The distance zero corresponds to the centre line of the excavation, and it is assumed that there is little or no gradient across the longwall face so that the subsidence profile is symmetrical about the excavation's centreline. In Fig. 4, it can be noticed that the empirical method provided by the Subsidence Engineers' Handbook (NCB 1975) fails to predict the correct depth of the trough, but it does predict the spatial extent very well. The results of the Mohr-Coulomb and Hoek-Brown failure criteria fail to predict both the depth and the spatial extent, but the results of both models agree closely with each other. This agreement means that the method of the calculation of the Mohr-Coulomb properties from the Hoek-Brown parameters is correct.

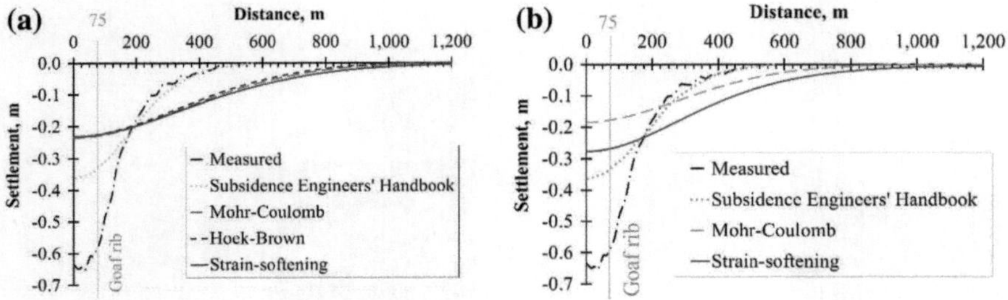

Fig. 4. Surface settlement half-profile, **a** constant GSI of 40, **b** increasing GSI

Further investigation includes the implementation of the strain-softening model. The model uses the Mohr-Coulomb failure criterion to detect failure and the cohesion of the rocks will suffer a post failure reduction in strength. Using test results, Pourhosseini and Shabanimashcool (2014) proved that the post failure friction angle is constant. For post-peak variations of inherent cohesion, Pourhosseini and Shabanimashcool (2014) suggested a function:

$$c = c_0\left(1 - \frac{\tanh(100\gamma_p)}{\tanh(10)} + 0.001\right)^n \tag{21}$$

where γ_p is the plastic shear strain, %; c_0 is the cohesion at the peak strength of the rock where $\gamma_p = 0$, and n is the fitting parameter, which depends on rock type and its magnitude varies from 0.29 for Sandstone to 0.34 for Mudstone (Pourhosseini and Shabanimashcool 2014). After the evaluation of the effect of this parameter on the subsidence profile, no effect was noticed after the goaf height was adjusted. A mean value of 0.3 was taken in this work.

Figure 4a demonstrates that the strain-softening constitutive model predicts an identical subsidence profile to those profiles obtained by the Mohr-Coulomb and Hoek-Brown models. However, the earlier investigations (i.e. by Lloyd et al. 1997) showed that the strain-softening model was capable of predicting a deeper trough than the Mohr-Coulomb model. This can be explained by examining the zone of plasticity. Figure 5a shows that the area of plastic deformation occurs directly above and under the excavation. After adjusting the goaf height, the strain-softening model shows the identical results as the Mohr-Coulomb model.

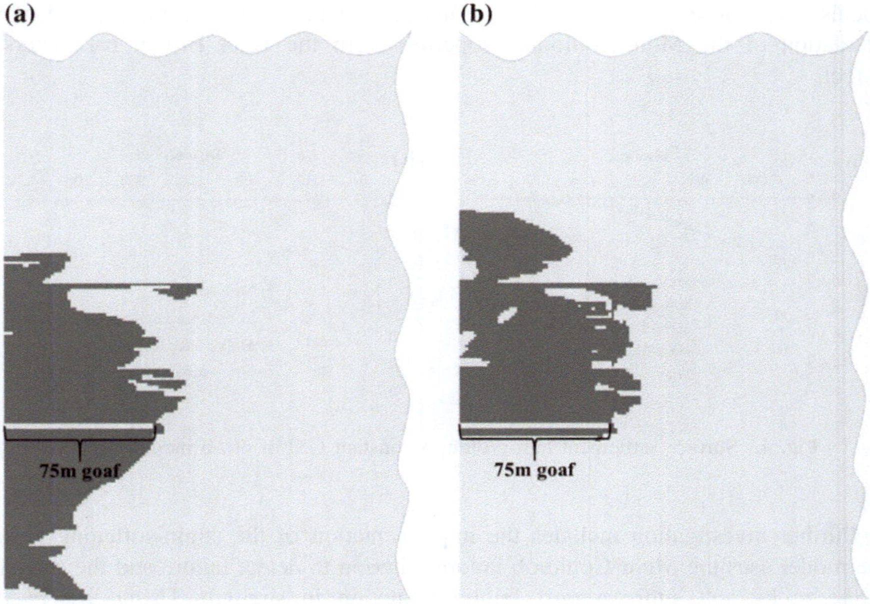

Fig. 5. Plastic deformation resulting from the excavation, **a** GSI is constant with the depth, **b** GSI gradually increased with the depth

4.2 Geological Strength Index

The choice of the GSI influences where the plastic deformations occur (above or under the goaf). As it was mentioned before, the GSI stands for Geological Strength Index, a system of rock-mass characterization. Practically, the GSI should increase with the depth because the deeper geomaterial is, the less weathered and in better condition it is. If the GSI increases with the depth, the failure area above the goaf appears abundant. To show this, the GSI was increased by 2 each under- and overburden layer from 25 to 85 from the surface to the bottom of the model as shown in Fig. 6a. The under- and overburden at hand is of multiple rock types including, mudstone, sandstone, seatearth,

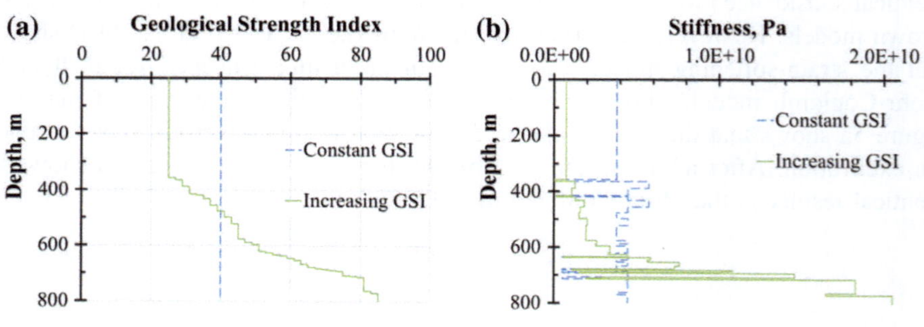

Fig. 6. **a** GSI with the depth, **b** stiffness vs depth

and siltstone (which is the roof material). The layers are too thin to be clearly shown on the diagram (Fig. 6). Figure 6a shows the constant GSI and the increasing GSI with the depth. Figure 6b presents the diagram of the stiffness, which is calculated according to Eq. 1 vs depth for two cases: when the GSI is constant and when the GSI changes from 25 to 85 with the depth according to Fig. 6a. In Fig. 5b, it can be seen that the model with the increasing GSI experiences more plastic defamation above the seam than the model with the constant GSI (Fig. 5a).

The plastic zone distribution influences the performance of the strain-softening model. In contrast to the model with the constant GSI (Fig. 4a), Fig. 4b shows that the strain-softening model with the increasing GSI predicts a deeper trough than the Mohr-Coulomb model; however, this trough predicts and erroneous width if it is compared to the field measurements. More research on computer modelling of the surface subsidence is needed.

4.3 Stresses in the Goaf

Based on earlier research, Derbin et al. (2016) concluded that the vertical stresses in the goaf after perturbation should recover to the natural stresses at some sufficient distance from the goaf rib; however, during the current simulation, it was noticed that this did not occur. Figure 7a presents both the theoretical stress distribution at a depth equal to the roof of the seam and several key characteristics, which help describe the stress arrangements after the collapse of a mine. They are three distances D1, D2, and D3; D1 is the distance into the goaf from the goaf rib where the stresses recover to the primary stress, D2 is the distance between the lowest residual stress in the goaf and the insitu stress in the unexcavated seem and D3 is the distance between the lowest stress in the goaf and the highest (peak) induced stress on the seam rib. D3 appears due to coal crushing at the seam rib, and it can be minimal or absent if crushing is not significant.

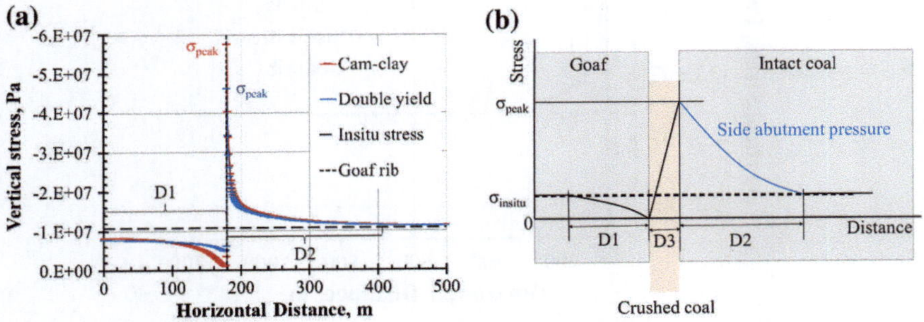

Fig. 7. Compression vertical stress within the goaf, **a** fictitious model (Derbin et al. 2016) and **b** theoretical (after Wilson 1983)

In order to investigate stress recovery in the goaf, Derbin et al. (2016) developed a simplified fictitious subsidence model where the goaf length was equal to distance D1. Distance D1 was calculated using the following equation suggested by Mukherjee et al. (1994) (after Wilson 1984):

$$D1 = 0.3 - 0.4H \tag{22}$$

where H = the thickness of the overburden.

The traditional double-yield model and the modified Cam-clay model were implemented to represent goaf material behaviour. Figure 7b shows that both models predict stresses of 6 MPa, which is lower than the primary stresses (10.5 MPa). It means that the both models fail to predict the correct behaviour of the goaf. At the same time, if Fig. 7a, b are collated, it can be noticed that the results obtained with the help of the modified Cam-clay model are closer to the theoretical expectations. The peak and lowest stresses predicted by the modified Cam-clay model are higher and lower respectively than the stresses predicted by the double-yield model. It means the modified Cam-clay model is better at predicting the behaviour of the goaf material.

Contrary to the research described above, where the goaf length is sufficient for recovering the primary stress, the goaf length is only 75 m in this research, which is more than three times less than necessary for the reestablishment of the primary stress. In the model for the Naburn site, H is 706.8 m, therefore following Eq. 22, D1 should be approximately 250 m. As a result, Fig. 8 shows that the stress in the goaf of the subsidence profiles obtained using the double-yield model and modified Cam-clay for the Naburn site are identical.

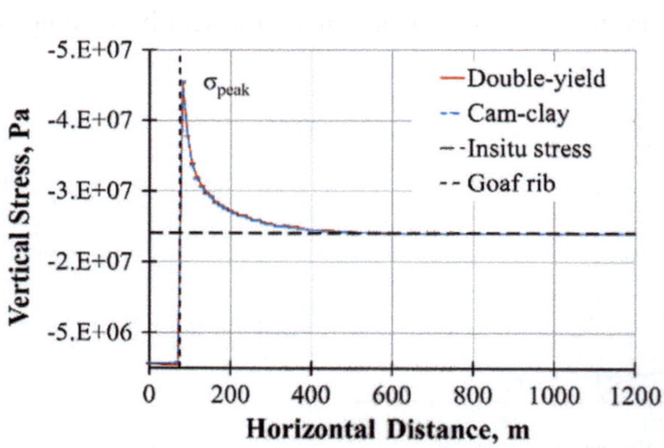

Fig. 8. Compression vertical stress within the goaf (Naburn model)

5 Conclusions

The paper describes how to simulate surface subsidence after a goaf collapse for any seam in any part of the world using only the description of the borehole log. The method of calculation of the Mohr-Coulomb properties out of the Hoek-Brown parameters was introduced. The predictions of the Mohr-Coulomb and modified Hoek-Brown constitutive models embedded in FLAC agree with each other. This proves the correctness of the method used to calculate the properties. The strain-softening method predicts identical deformation, magnitude, and spatial extent for a constant GSI, but the variable GSI causes a deeper trough.

A pattern of plastic deformation around the goaf is crucial for the strain-softening model. When plastic deformation above and under the goaf is in the same quantity, the strain-softening effect of the model is compensated by adjusting the goaf height. All these plastic zones act as one big goaf. The pattern of plastic deformation depends on relationship between the GSI and depth. The GSI increasing with the depth produces a larger failure area above the goaf, which deepens the trough of the strain-softening model. Unfortunately, this is not sufficient to match field observations in this research, and the trough keeps the same erroneous width. Hence, practitioners should use traditional constitutive models to predict surface subsidence with great care.

For the time being, the best solution would be utilization of the strain-softening constitutive model keeping in mind that it can underestimate the depth and overestimate the width of the subsidence trough. To improve the prediction of the sizes of the trough, more advanced constitutive models should be implemented. The recent research by Derbin et al. (2018) has shown that the bubble model has better results. It predicts a deeper and narrower trough than the strain-softening model does.

The paper also discusses how to improve the modelling of the goaf behaviour. A Critical State model, i.e. the modified Cam-clay model, was implemented into the goaf. It was shown that if the goaf length is not sufficient to recover the primary stresses, the modified Cam-clay model and the double-yield model predict identical goaf behaviour. Hence, it is difficult to say how much of the model discrepancies can be attributed to the goaf and how much can be attributed to the 700 m of overburden. This work suggests deeper investigation of modelling the goaf behaviour by increasing the goaf length up to distance D1 and implementing more advanced constitutive models into the goaf.

Acknowledgements. The authors greatly acknowledge the open fund SKLGDUEK1512 of the China University of Mining and Technology.

References

Itasca.: Manual FLAC3D Version 5.01 (2013)

Badr, S., Ozbay, U., Kieffer, S., Salamon, M.: Three-dimensional strain softening modeling of deep longwall coal mine layouts. FLAC Numer. Model. Geomech. 233–239 (2003). https://doi.org/10.1201/9780203883204.ch79

Balmer, G.A.: General analytic solution for Mohr envelope. In: Proceedings-American Society for Testing and Materials, pp. 1260–1271 (1952)

Bell, F.G., Donnelly, L.J.: Mining and its Impact on the Environment, p. 547. Taylor & Francis (2006). https://doi.org/10.2113/gseegeosci.13.3.270

Bieniawski, Z.T.: Rock mechanics design in mining and tunneling. Balkema. (1984)

Bräuner, G.: Subsidence due to underground mining: theory and practices in predicting surface deformation, US Bureau of Mines, p. 8571 (1973). https://doi.org/10.1016/0148-9062(74)91643-x

British Standard.: Code of practice for site investigation, p. 147 (1981)

Deere, D.: Geological considerations. Rock mechanics in engineering practice, pp. 1–20 (1968)

Derbin, Y., Walker, J., Wanatowski, D., Marshall, A.: Issues related to goaf modeling. In: Proceedings of the 19th Southeast Asian Geotechnical Conference & 2nd AGSSEA Conference, pp. 1015–1021 (2016)

Derbin, Y., Walker, J., Wanatowski., Marshall, A.: Implementation of advanced constitutive models for the prediction of surface subsidence after underground mineral extraction. (Expected to be published) (2018)

Gens, A., Potts, D.: Critical state models in computational geomechanics. Eng. Comput. **5**, 178–197 (1988). https://doi.org/10.1016/0148-9062(89)90209-x

Hansen, T.H.: Rock properties. Norwegian Rock and Soil Assoc. **5**, 41–44 (1988)

Herrero, C., Munoz, A., Catalina, J.C., Hadj-Hassen, F., Kuchenbecker, R., Spreckels, V., Juzwa, J., Bennet, S., Purvis, M., Bigby, D., Moore, D.: Prediction and monitoring of subsidence hazards above coal mines (Presidence), p. 138. Directorate-General for Research and Innovation, Luxembourg (2012)

Hoek, E., Brown, E.T.: Underground excavations in rock, p. 527. Institution of Mining and Metallurgy, London (1980)

Hoek, E., Brown, E.T.: Practical estimates of rock mass strength. Int. J. Rock Mech. Min. Sci. **34**, 1165–1186 (1997). https://doi.org/10.1016/s0148-9062(97)00305-7

Hoek, E., Diederichs, M.S.: Empirical estimation of rock mass modulus. Int. J. Rock Mech. Min. Sci. **43**, 203–215 (2006). https://doi.org/10.1016/j.ijrmms.2005.06.005

Hoek, E., Wood, D., Shah, S.: A modifed Hoek-Brown failure criterion for jointed rock masses. In: Rock Characterization: ISRM Symposium, Eurock'92, Chester, UK, 14–17 September 1992, pp. 209–214 (1992)

Hudson, J.A., Harrison, J.P.: Engineering rock mechanics: an introduction to the principles, p. 444, Elsevier (2000). https://doi.org/10.1115/1.1451165

Indraratna, B., Salim, W.: Modelling of particle breakage of coarse aggregates incorporating strength and dilatancy. Proc. Inst. Civil Eng. London **155**(4), 243–252 (2002). https://doi.org/10.1680/geng.155.4.243.38691

ISRM: Commission on standardisation laboratory and field results. Suggested methods for determining hardness and abrasiveness of rocks. Int. J. Rock Mech. Min. Geomech. Abstr. **15**, 89–97 (1988)

Lama, R., Vutukuri, V.: Handbook on mechanical properties of rocks-testing techniques and results. **3** (1978)

Lloyd, P., Mohammad, N., Reddish, D.: Surface subsidence prediction techniques for UK coalfields-an innovative numerical modelling approach. In: Proceeding of the 15th Mining Congress of Turkey, Ankara, pp. 6–9 (1997)

Mukherjee, C., Sheorey, P.R., Sharma, K.G.: Numerical simulation of caved goaf behaviour in longwall workings. Int. J. Rock Mech. Min. Sci. Geomech. **31**, 35–45 (1994). https://doi.org/10.1016/0148-9062(94)91292-0

Najafi, M., Jalali, S.M.E., Khalokakaie, R.: Thermal–mechanical–numerical analysis of stress distribution in the vicinity of underground coal gasification (UCG) panels. Int. J. Coal Geol. **134–135**, 1–16 (2014). https://doi.org/10.1016/j.coal.2014.09.014

NCB.: National Coal Board. Subsidence Engineers' Handbook, p. 111. Mining Department, London (1975)

Palmström, A.: A Rock Mass Characterization System for Rock Engineering Purpose, p. 381. University of Oslo (1995)

Palmström, A., Singh, R.: The deformation modulus of rock masses—comparisons between in situ tests and indirect estimates. Tunn. Undergr. Space Technol. **16**, 115–131 (2001). https://doi.org/10.1016/s0886-7798(01)00038-4

Pourhosseini, O., Shabanimashcool, M.: Development of an elasto-plastic constitutive model for intact rocks. Int. J. Rock Mech. Min. Sci. **66**, 1–12 (2014). https://doi.org/10.1016/j.ijrmms.2013.11.010

Salamon, M.: Mechanism of caving in longwall coal mining. In: Rock Mechanics Contributions and Challenges: Proceedings of the 31st US Symposium, pp. 161–168. Golden, Colorado (1990)

Salamon, M.: Rockburst hazard and the fight for its alleviation in South African gold mines. Rockbursts: prediction and control, pp. 11–36, IMM, London (1983). https://doi.org/10.1016/0148-9062(84)90579-5

Shtumpf, G.G.: Mechanical properties of rocks of the Kuznetsk basin and regularities of their change. Phys. Chem. Probl. Min. **4**, 43–51 (1994). (in Russian)

Singh, G.S.P., Singh, U.K.: Assessment of goaf characteristics and compaction in longwall caving. Min. Technol. **120**, 222–232 (2011). https://doi.org/10.1179/1743286311y.0000000010

Wilson, A.H.: The stability of underground workings in the soft rocks of the coal measures. Int. J. Min. Eng. **1**, 91–187 (1983). https://doi.org/10.1007/bf00880785

Wilson, A.H.: Pillar stability in longwall mining. In: Wilson, A.H. (ed.) State-of-the-Art of Ground Control in Longwall Mining and Mining Subsidence, pp. 85–95 (1984). https://doi.org/10.1016/0148-9062(84)90575-8

Xiao, Y., Sun, Y., Liu, H., Yin, F.: Critical state behaviors of a coarse granular soil under generalized stress conditions. Granular Matter **18**(2), 17 (2016). https://doi.org/10.1007/s10035-016-0623-3

Xu, N., Kulatilake, P.H.S.W., Tian, H., Wu, X., Nan, Y., Wei, T.: Surface subsidence prediction for the Wutong mine using a 3-D finite difference method. Comput. Geotech. **48**, 134–145 (2013). https://doi.org/10.1016/j.compgeo.2012.09.014

A New Method Based on PFC3D and Hierarchical Modeling for Pile Foundation Analysis

Honghua Zhao(✉), Jin Zhang, Peng Qiu, and Shunying Ji

State Key Laboratory of Structural Analysis for Industrial Equipment, Department of Engineering Mechanics, International Research Center for Computational Mechanics, Dalian University of Technology, Dalian 116023, China
zhaoh@dlut.edu.cn

Abstract. By developing a hierarchical modeling method, using discrete element analysis software PFC3D, a pile foundation model is built. According to the idea of finite element method, using smaller particles for concerned part and use larger particles for the part which is close to the model boundary. By adjusting the micromechanical parameters of the particles, the macroscopic mechanical properties of the soils constituted by different particle diameters are kept constant, which is verified by the triaxial test. This method can be applied to study the soil-structures interaction problems with DEM method by saving computation cost.

Keywords: Discrete element method · Pile foundation · Hierarchical modeling

1 Introduction

Deep foundation is a type of foundation distinguished from shallow foundations by the depth they are embedded into the ground. There are different terms used to describe different types of deep foundations including the pile, the pier, drilled shafts, and caissons. Its role is to transfer the load which is relatively concentrated to the deep foundation. Pile foundation is the most widely used deep foundation in the world today. They are used in high buildings, harbor, bridges etc. It is typically connected with a pile cap at the pile head. The study of pile–soil–cap interaction is important to the design of pile foundation and to get insights into the load transfer mechanisms. Hongladaromp et al. [1] have investigated a capped pile group, but did not consider surface interactions of pile-pile and pile-soil. Kuwabara [2] used a boundary element method to analyze the behaviour of a capped pile group. Clancy and Randolph [3] and Poulos [4] presented an approximate numerical solution mainly based on finite element theory and Mindlin's solution, respectively. Ta and Small [5, 6] have analyzed piled raft systems subjected to vertical loads in layered soil using a proposed finite layer theory. Zhang and Small [7] analyzed the capped pile groups subjected to horizontal and vertical loads, in which finite element method is used to analyze the cap and piles while the finite layer theory is employed to analyze the layered soil. Reul [8]

© Springer International Publishing AG, part of Springer Nature 2019
A. Sevi et al. (eds.), *Enhancements in Applied Geomechanics, Mining, and Excavation Simulation and Analysis,* Sustainable Civil Infrastructures,
https://doi.org/10.1007/978-3-319-95645-9_10

investigated the bearing behavior of piled rafts in overconsolidated clay and concluded that the interaction between piles and raft is a major influence. Sinha and Hanna [9] have carried a 3-D numerical analysis on piled raft foundation also using finite element method.

The numerical methods can be categorized into two main categories: continuum and discontinuum approaches [10]. After many years' development, finite element method has become very mature and widely applied in geotechnical engineering field [11, 12]. Compared with the discrete element method, its computational efficiency is a great advantage. However, finite element method which based on continuum mechanics could not describe the microstructure accurately of the soil and provide the micro mechanics insights into the load transfer mechanism [13, 14]. Meanwhile, soil is discontinuous on itself, when the finite element method is implemented, the hypothesis brought into the simulation will affect the accuracy of the result. Cundall and Strack first developed the discrete element method (DEM) [15, 16], which is currently an effective simulation method to obtain inter-particle forces and trace microstructure evolution of granular materials. Many researchers have conducted DEM simulations on the soil behavior around pile foundations. erface undergoing loading and stress relief is studied [17], and soil-pile interaction during pile penetration is also investigated [18]. Liu and Zhou [19] simulated the pile-soil interaction subject to uplift load. Lobo-Guerrero and Vallejo [20] investigated the influences of pile shape and pile interaction on the crushable behavior of granular materials around driven piles. A discrete element modeling of a granular platform supported by piles in soft soil is performed to validate a small scale model test [21]. Lai et al. [22] studied the "soil-arching" within geogrid-reinforced and unreinforced pile-supported embankments with DEM. Ng and Meyers [23] investigated the side resistance of drilled shafts in granular soils by DEM. Because the computation cost is very high by DEM, the combined DEM and other continuous numerical methods has been developed and applied to geomechanics problems, such as coupled finite element and discrete element method [24, 25], coupled discrete element–finite difference method [26] etc. However these coupled methods are more complicated when performing the analysis.

At present, it is difficult to build the model at the real scale for engineering problems with DEM method. Since the sand particle size is very small, to closely simulate the real conditions, millions of particles are needed to build a pile-soil model for running a DEM simulation, which is difficult to realize now based on personal computers. In the current study, in order to obtain micro mechanics behavior and also save computation time, a radial particle expansion method called hierarchical modeling is proposed to study the pile-soil-cap interaction behavior for non-displacement piles.

2 DEM Model and Material Properties

In discrete element modeling, the soil is modeled as an assembly of interacting particles. Different constitutive models are used to define interaction forces between particles. Newton's second law is used to compute the movement of the particles.

A discrete element modeling allows a better understanding of the phenomenon at the particle scale governing the macroscopic behavior of the system.

2.1 Hierarchical Modeling Method

It is proposed here that a discrete element model can be divided into several concentric annular regions with different particle diameter, similar as that commonly used mesh generation in finite element modeling, called particle radial expansion method. The interested central research zone is modeled with smaller sized particles and the particle's diameter increases with the distance increasing away from the central zone. In application, the modeling zone is divided into several different zones with the same porosity. If the macro behavior of each zone is consistent with each other, they can be used to model the same soils. In order to obtain the same macro behavior, different modeling parameters can be assigned to the particles in different zones. For the pile–soil–cap model in this study, the model is divided into five zones. Zones next to each other have the continuous variation of particle diameter, which means the smallest particle size is going to be the largest particle size in next neighboring zone. This avoids the effects of the sudden change in grain sizes. Four zones with the same macro behavior, a boundary zone does not need to have the same macro behavior and it can be assigned with the same micro parameter as the fourth zone. The boundary zone functions as a zone to absorb the stress transferred to the boundary and reduce the boundary effect of walls. Based on the Saint Venant's Principle, the force effects are localized and the boundary layer will not cause difference for the central zone.

A 10 m × 10 m × 2.5 m soil mass model was built in PFC3D 3.10 as shown in Fig. 1. How to assign the appropriate micro parameters for the soils of different zones needs to be solved first in this study. A series of triaxial test were conducted to find the parameters which can provide the same stress-strain relationships for the soils in different zones. Triaxial tests were conducted with different confining pressures of 100, 150, and 200 kPa. All the triaxial testing samples have the same porosity of 0.4.

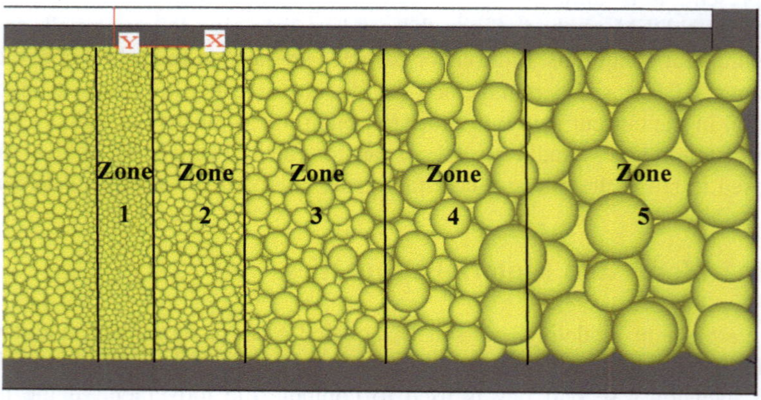

Fig. 1. Soil models generated with radial particle expansion method

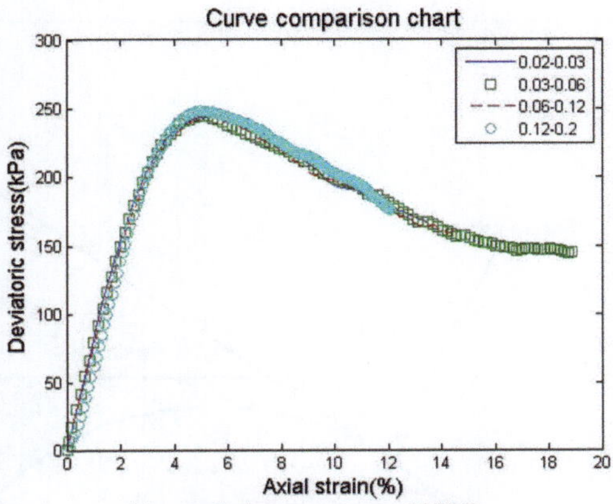

(a) confining pressure 100kPa

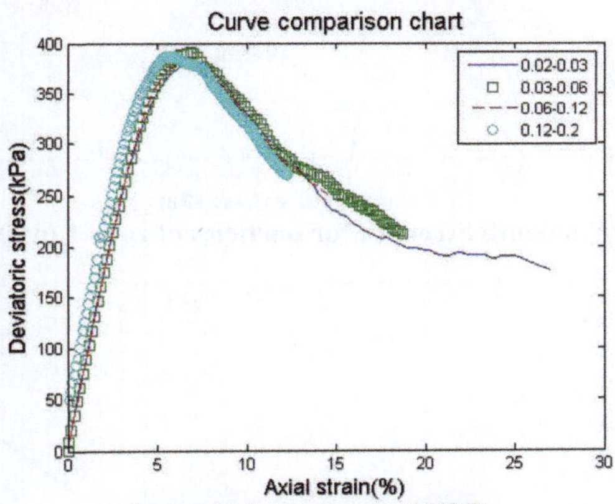

(b) confining pressure 150kPa

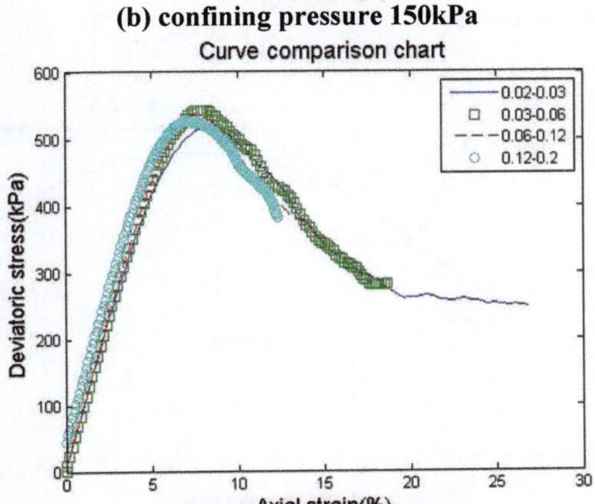

(c) confining pressure 200kPa

Fig. 2. Stress-strain curves for soils of different zones under confining pressure of 100, 150, and 200 kPa

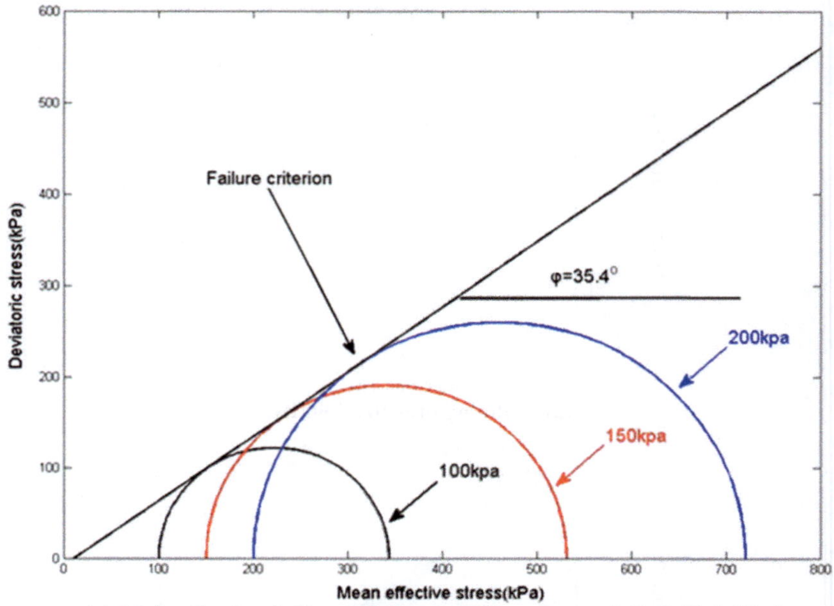

(a) Mohr-Coulomb Strength for particles of zone 1 (d=0.02-0.03m)

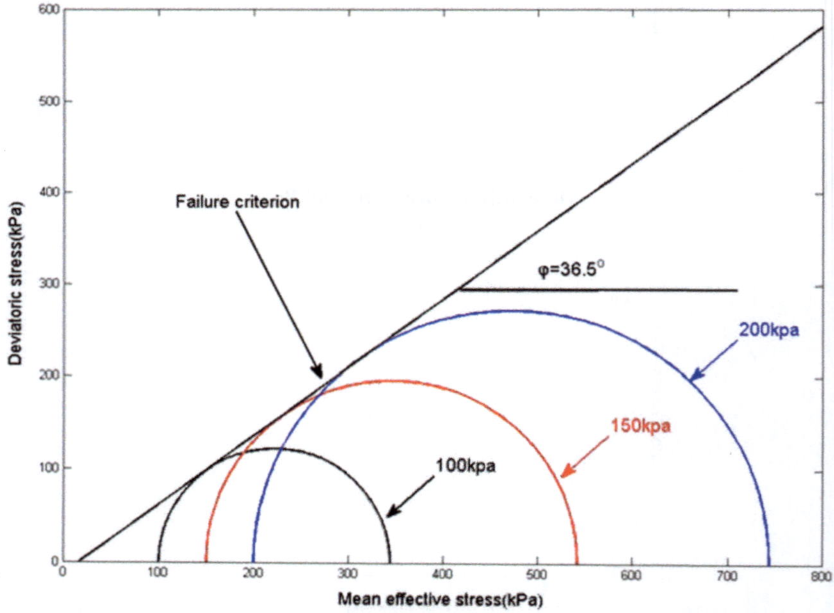

(b) Mohr-Coulomb Strength for particles of zone 2 (d=0.03-0.06m)

Fig. 3. Determination of the internal friction angle by the use of Mohr circles

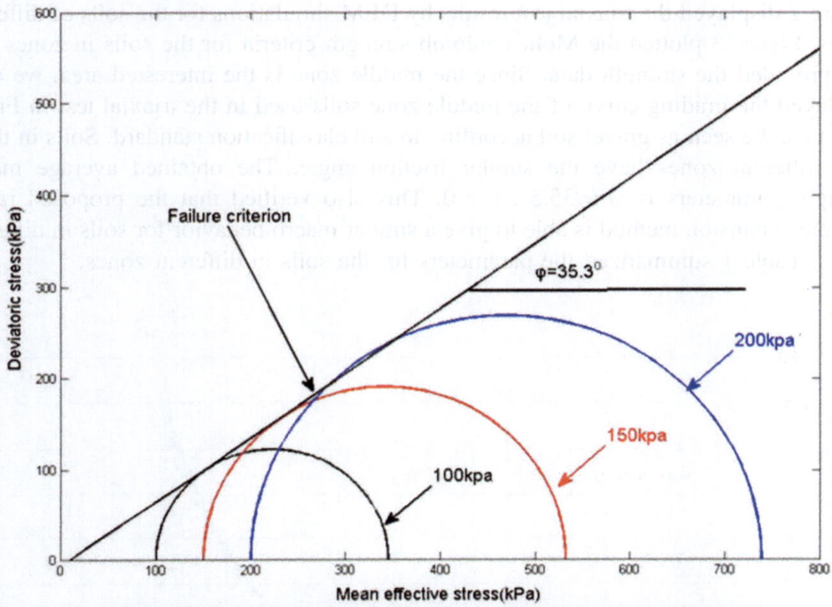

(c) Mohr-Coulomb Strength for particles of zone 3(d=0.06-0.12)

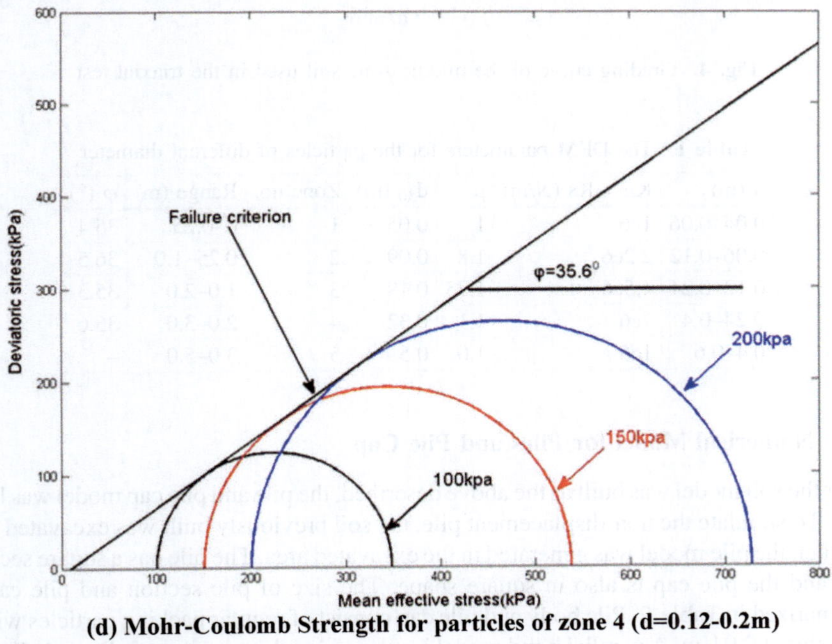

(d) Mohr-Coulomb Strength for particles of zone 4 (d=0.12-0.2m)

Fig. 3. (*continued*)

Figure 2 displayed the triaxial test results by DEM simulations for the soils of different zones. Figure 3 plotted the Mohr coulomb strength criteria for the soils in zones 1–4 and provided the strength data. Since the middle zone is the interested area, we only displayed the grading curve of the middle zone soils used in the triaxial test in Fig. 4 and it can be seen as gravel soil according to soil classification standard. Soils in these four different zones have the similar friction angle. The obtained average macro strength parameters is $\varphi = 35.5°$, c = 0. This also verified that the proposed radial particle expansion method is able to give a similar macro behavior for soils in different zones. Table 1 summarized the parameters for the soils in different zones.

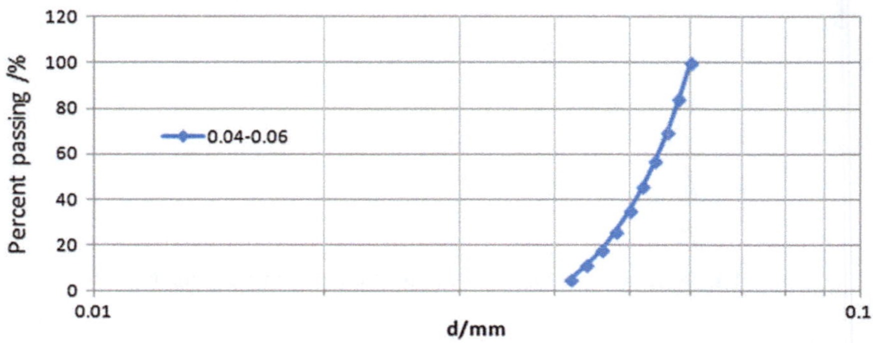

Fig. 4. Grading curve of the middle zone soil used in the triaxial test

Table 1. The DEM parameters for the particles of different diameter

d (m)	Kn = Ks (N/m)	μ	d_{50} (m)	Zone no.	Range (m)	φ (°)
0.04–0.06	1e6	1	0.05	1	0–0.25	35.4
0.06–0.12	2.2e6	1.8	0.09	2	0.25–1.0	36.5
0.12–0.24	4.5e6	1.75	0.18	3	1.0–2.0	35.3
0.24–0.4	7e6	1.1	0.32	4	2.0–3.0	35.6
0.4–0.6	1e6	1.0	0.5	5	3.0–5.0	–

2.2 Numerical Model for Piles and Pile Cap

After the soil model was built as the above described, the pile and pile cap model was built next. To simulate the non-displacement pile, the soil previously built was excavated first and then the pile model was generated in the excavated area. The pile has a square section area and the pile cap is also in square shape. The size of pile section and pile cap is summarized in Table 2. Pile body and pile cap consist of regular packing particles with a diameter of 0.01 m. A parallel bond model is assigned to the particles of pile and pile cap with the following model parameters: $p_{b}\_rad == 1.5$, $p_{b}\_k_n = 1.0 \times 1010$ N/m^3, $p_{b}\_k_s = 1.0 \times 1010$ N/m^3, $p_{b}\_sstren = 1.0 \times 1010$ N/m^2, $p_{b}\_nstren = 1.0 \times 1010$ N/m^2. The available model parameters are listed in Table 2. The maximum ratio of soil model's width to the pile cross section length is 22.72 greater than 20, which minimized the boundary effect. To make the study simple, total 8 piles were built as listed in Table 3. Model 0MM is a single pile without cap. Model LMM, MMM, SMM have different pile cap size but with the same pile length and pile section area. Model MMM, MML, MMS have the same pile cap size and pile section but different pile length. MBM, MSM, MMM have different pile diameter with the same pile cap size and pile length. Pile model MMM was used in different group studies. All the pile models are shown in Fig. 5.

Table 2. Basic physical parameters of pile and cap

Particle density (kg/m^3)	Particle radius (m)	K_n (N/m)	K_s (N/m)	μ
3000	0.01	5×10^6	1×10^6	1.2
$P_b\_k_n$ (N/m^3)	**$P_b\_k_s$ (N/m^3)**	**σ_n (N/m^2)**	**τ_n (N/m^2)**	**R_b**
10^{10}	10^{10}	10^{10}	10^{10}	1.5

Table 3. Size of single pile model with and without cap

Pile	Cap side length B (m)	Pile length L (m)	Pile section length d (m)	L/B	B/d	L/d
LMM	0.62	1.4	0.14	2.258	4.428	10
MMM	0.46	1.4	0.14	3.043	3.285	10
SMM	0.3	1.4	0.14	4.666	2.142	10
0MM	–	1.4	0.14	–	–	10
MML	0.46	1.8	0.14	3.913	3.285	12.85
MMS	0.46	1.0	0.14	2.173	3.285	7.142
MLM	0.46	1.4	0.22	3.043	2.091	6.363
MSM	0.46	1.4	0.1	3.043	4.6	14

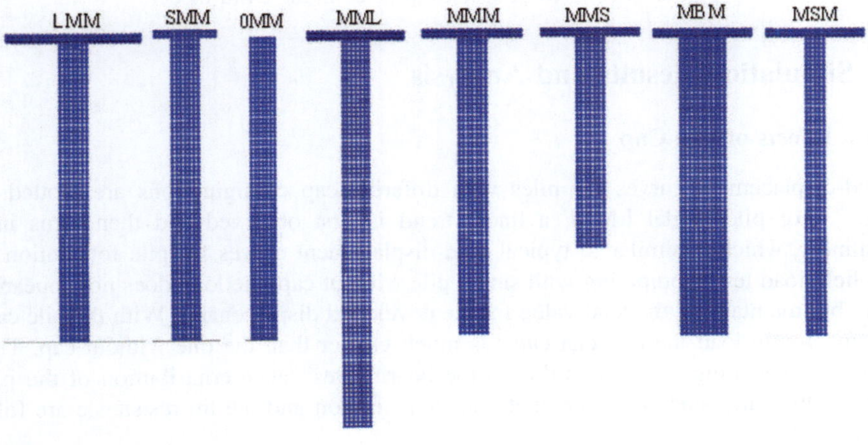

Fig. 5. Model series of single pile with and without pile cap

In order to trace the change of porosity and stress of sand particles during loading, a series of measurement spheres are installed in the model as shown in Fig. 6. Because of the symmetry of the model, measurement spheres are installed only on the right half of the model. More measurement spheres are installed below the pile bottom and around the pile cap. Since the stress is the average stress over the sphere, to make the measured stress more uniform and reliable, overlapping of measurement spheres is allowed. The measured quantity is the average over the volume of measurement sphere. Total 132 measurement spheres are installed.

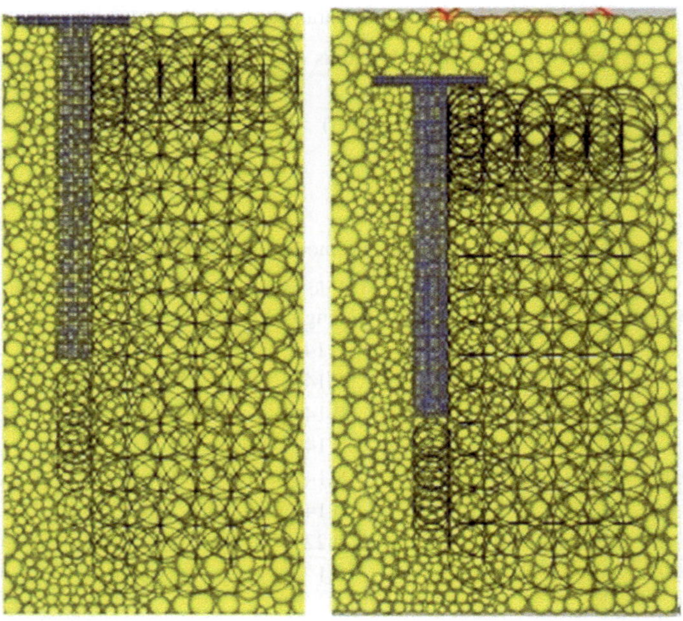

Fig. 6. Measurement spheres installed in the DEM model

3 Simulation Results and Analysis

3.1 Effects of Pile Cap

Load-displacement curves for piles with different cap configurations are plotted in Fig. 7. For pile model LMM, a linear trend is first observed and then turns into nonlinear, which is similar to typical load-displacement curves for pile foundation in the field load test. Comparing with single pile without cap, the load does not appear to drop but maintaining the load value for the developed displacement. With the pile cap, the maximum load the pile can carry is much greater than the one without cap. This increase of loading capacity is due to the bearing resistance contribution of the pile cap. At the early stage of settlement, the shaft friction and the tip resistance are fully developed.

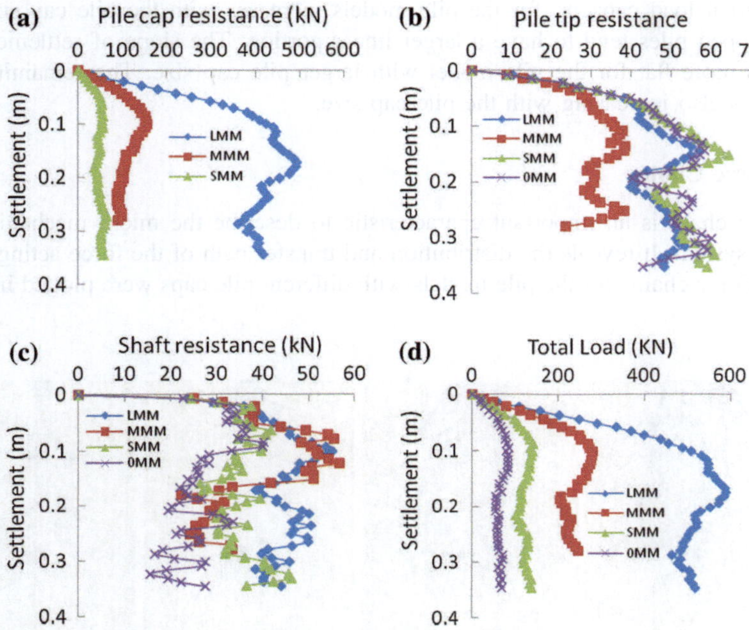

Fig. 7. The variation of cap resistance, tip resistance, shaft resistance, and total load for piles with different pile caps

The pile cap resistance increases with the pile cap size. Pile cap resistance tends to increase first nonlinearly to a maximum and then drops gradually to a residual state. This can be explained that when the load applied exceeds the bearing capacity of the soil beneath the pile cap, the soil fails finally. With the larger pile cap size, the reduction of pile cap resistance tends to be more significant.

For pile model 0MM, the shaft friction first increases to the maximum value then drops to a certain constant value. This indicates that the static friction between pile and soil contributes to the shaft friction at the beginning; once the pile is moving, the static friction turns into smaller sliding friction. Pile model SMM also has a similar variation but with a slightly larger shaft resistance value than 0MM. Pile model MMM and LMM have an increasing shaft resistance at the initial settlement and then drop sharply. After the drop, the shaft resistance increases back again to a certain value. Pile model LMM has a larger shaft resistance when the settlement is larger than 0.18 m while MMM has a smaller value of shaft resistance.

The pile tip resistance for all the pile models fully develops at the early settlement stage and then drops a little bit, after that gradually increases. The soil surrounding the pile tip started to move with loading which reduces the bearing resistance of soil and caused the first drop of pile tip resistance. Fully developed shaft friction happens earlier than the pile tip resistance. Model MMM has the lowest pile tip resistance, while the other three models have a similar vale of pile tip resistance.

The total load capacity for the pile models increases with the pile cap size. The larger capped piles tend to have a larger linear portion. The slope of settlement-load curves is more flat for the pile model with larger pile cap size. The remaining load capacity is also increasing with the pile cap size.

3.2 Force Chains

The force chain is an important characteristic to describe the micro mechanics of a granular system. It reveals the distribution and transfer path of the force acting on the system. Force chains for the pile models with different pile caps were plotted in Fig. 8

Fig. 8. Soil force chain distribution with different cap size

at different settlement s = 0.05, 0.137, and 0.288 m. Pile model LMM has a very complex force chain net. But obviously the strong force chains were developed below the pile cap on both sides of the pile body. These force chains transfers the load mostly downward with some portion away from the pile. There are weak chains along the lower part of the pile above the pile bottom. Below the pile bottom, there is some force chains also formed. These force chains are not as strong as those below the cap but stronger than those weak force chains along the lower part of the pile. For the pile with small pile cap or without pile cap, the force chains are weak and the major force chains only located below the pile bottom. Arching effect was observed for LMM and MMM because of the strong force chains formed on the top part of the model. These characteristics of force chains might be able to help us to design better pile foundation systems.

3.3 Stress Field

The stress status was monitored through the installed measurement spheres. The stress field of σ_z on the central plane was plotted in Fig. 8 at three different settlement values of s = 0.05, 0.137, and 0.228 m. Here the negative value of σ_z means compressive stress through the paper. First, σ_z is only mobilized in a cone shape for pile model 0MM, while for the capped pile, it is in a shape like "R". σ_z is distributed in a large zone from the top to the bottom of the capped pile, the larger the pile cap and the greater the mobilized soil zone. While for no capped pile, it is mainly distributed in the lower part. Two high stress distribution zones are observed for the capped pile, one is below the pile cap, and the other is below the pile bottom. σ_z around the pile bottom is not significantly affected by the pile cap. The mobilized soil zone is expanding with the settlement increasing as shown in Fig. 9. This explains that the effect of pile cap is to mobilize more soil to take the load from the pile.

3.4 Porosity

From the DEM simulation, through the installed measurement spheres, porosity distribution can be obtained. Porosity is an important mesoscopic parameter to describe the granular materials. Porosity distribution on the central plane of the model was plotted in Fig. 10 at three different settlement values of 0.05, 0.137, and 0.228 m. Three distinct porosity zones are observed from the plots. A zone with larger soil porosity is next to the pile because of the excavation, follows a second narrow smaller porosity zone, then a third zone with a porosity value among the above two zones. As observed from these plots, there is a relatively large gradient change for the porosity close to the pile. Further away from the pile, there is almost no change in porosity. With the pile moving down, the soil is compacted below the pile cap and the porosity is reducing, while the soil around the lower pile body is increasing. The porosity of soil below the pile bottom is reducing due to settlement of the pile tip. This phenomenon may be due to the dragging effects of the pile. The pile is moving down, while the dilatancy effects made the sand particles around the pile loose. These results also revealed that the porosity change is limited to a narrow zone close to the drilled piers.

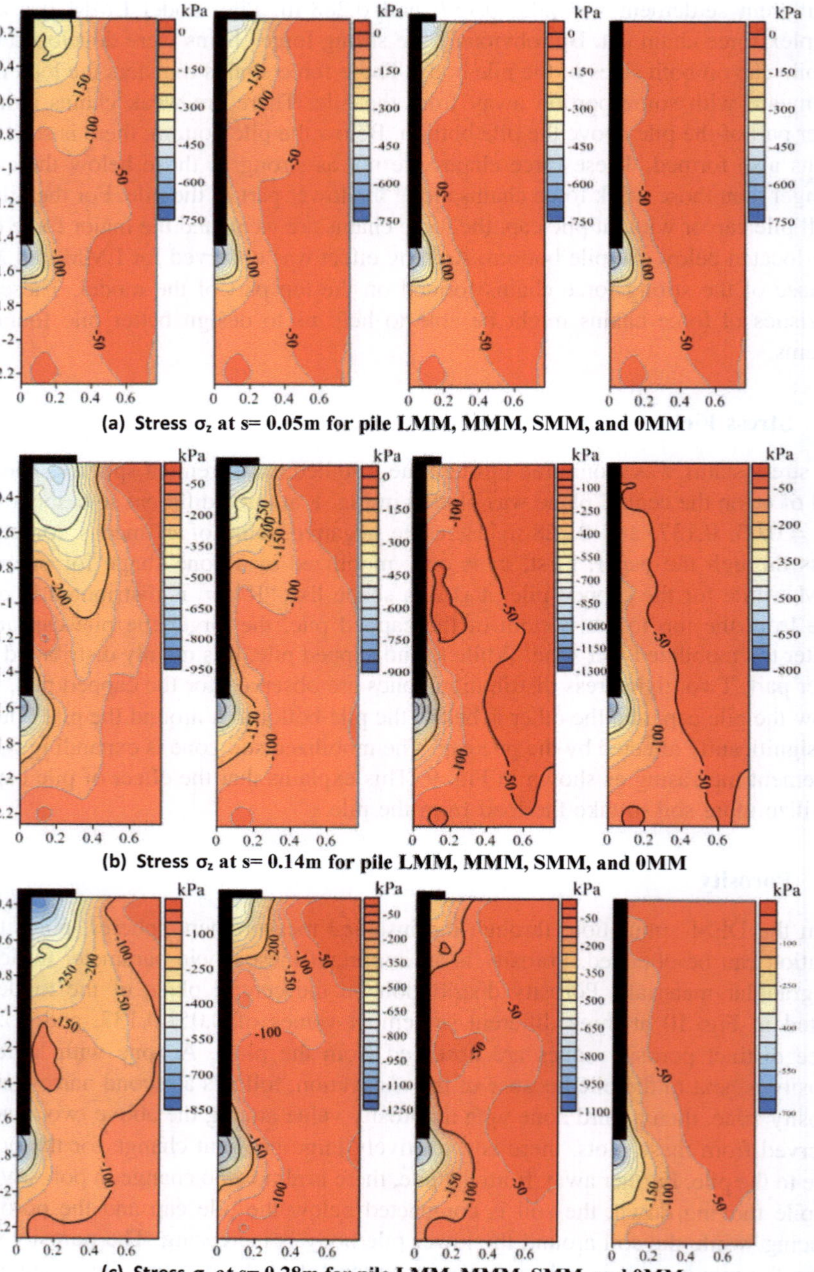

Fig. 9. Stress distribution of soil mass with different cap size

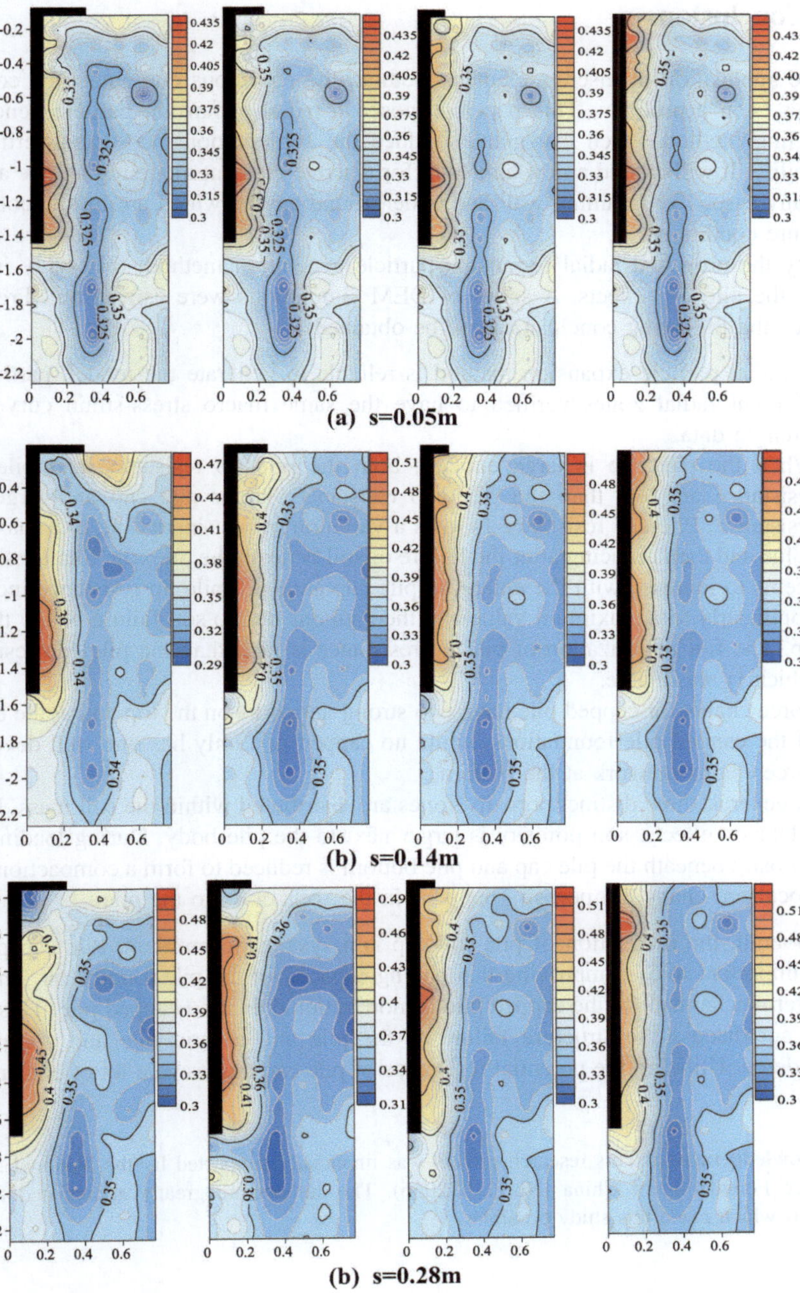

Fig. 10. Soil porosity distribution map with different pile cap size

4 Conclusions

In the application of discrete element method, the enormous computational cost has always been a problem. This paper proposed a radial expansion particle generation modeling method which can greatly reduce the computational cost and verified its reliability. It will extend a new approach for discrete element modeling and be applied to many cases like retaining wall and underground cavern which are in low confining pressure conditions.

By the proposed radial expansion particle generation method, this study investigated the pile cap effects. A series of DEM simulations were carried out. From the results, the following conclusions can be obtained:

1. A radial particle expansion method is reliable to generate the model, particles in different radial zones verified to have the same macro stress-strain curves and strength data.
2. When the pile cap is large enough, both the pile tip resistance and pile shaft resistance are less than the pile cap resistance, larger pile cap and larger pile resistance. Pile cap resistance follows a variation of first increasing to a maximum value and then reducing after the failure of soil beneath the pile cap. Shaft resistance seems to increase with the enlarging pile cap area. All pile tip resistance increases nonlinearly to a maximum value and then softens due to soil failure below the pile tip. The full mobilization of pile tip resistance is later than the pile cap resistance which is reasonable.
3. Force chains for capped pile have two strong networks on the top and at the bottom of the capped pile foundations, while no capped pile only has one well developed force chains network at the bottom.
4. In general, three distinct porosity zones are distributed within the soil mass. Due to dilatancy effects, soil porosity is larger next to the pile body. During loading, soil porosity beneath the pile cap and pile bottom is reduced to form a compaction zone. Localized change in porosity happens in the soils close to the pile.

Overall, the interaction of pile-soil-cap is a systematic problem. Pile cap plays a very important role in improving the loading capacity of the pile foundation. The load transfer mechanism for the capped pile is more rational than no capped pile. This paper does not discuss the structure failure of the pile and a rigid pile and a rigid cap considered. Utilizing the potential resistance of pile cap may save a lot of construction cost.

Acknowledgements. This research project was financially supported by the National Natural Science Foundation of China (No. 11672066). The authors are greatly appreciated for this support which made this study possible.

References

1. Hongladaromp, T., Chen, N.J., Lee, S.L.: Load distribution in rectangular footings on piles. Geotech. Eng. **4**(2), 77–90 (1973)
2. Kuwabara, F.: An elastic analysis for piled raft foundations in a homogeneous soil. Soils Found. **29**(1), 82–92 (1989)
3. Clancy, P., Randolph, M.F.: An approximate analysis procedure for piled raft foundations. Int. J. Numer. Anal. Meth. Geomech. **17**, 849–869 (1993)
4. Poulos, H.G.: An approximate numerical analysis of pile–raft interaction. Int. J. Numer. Anal. Meth. Geomech. **18**, 73–92 (1994)
5. Ta, L.D., Small, J.C.: Analysis of piled raft systems in layered soils. Int. J. Num. Anal. Methods Geomech. **20**, 57–72 (1996)
6. Ta, L.D., Small, J.C.: An approximation for analysis of raft and piled raft foundations. Comput. Geotech. **20**(2), 105–123 (1997)
7. Zhang, H.H., Small, J.C.: Analysis of capped pile groups subjected to horizontal and vertical loads. Comput. Geotech. **26**, 1–21 (2000)
8. Reul, O.: Numerical study of the bearing behavior of piled rafts. Int. J. Geomech. **2**(4), 59–68 (2004). https://doi.org/10.1061/(asce)1532-3641
9. Sinha, A., Hanna, A.M.: 3D numerical mode for piled raft foundation. Int. J. Geomech. **17** (2), 04016055 (2017)
10. Duan, K., Kwok, C.Y., Tham, L.G.: Micromechanical analysis of the failure process of brittle rock [J]. Int. J. Numer. Anal. Meth. Geomech. **39**(6), 618–634 (2015)
11. Gourvenc, S.M., Powrie, W.: Three-dimensional finite element analysis of diaphragm wall installation. Geotechnique **49**(6), 801–823 (1999)
12. Yasushi, A., Osamu, K., Osamu, M., et al.: A numerical study of ground displacement and stress during and after the installation of deep circular diaphragm walls and soil excavation. Comput. Geotech. **35**(5), 791–807 (2008)
13. Cundall, P.: A discontinuous future for numerical modelling in geomechanics? Proc. ICE Geotech. Eng. **149**(1), 41–47 (2001)
14. Potyondy, D.O., Cundall, P.A.: A bonded-particle model for rock. Int. J. Rock Mech. Min. Sci. **41**(8), 1329–1364 (2004)
15. Cundall, P.A., Strack, O.D.L.: A discrete numerical model for granular assemblies. Geotechnique **29**, 47–65 (1979)
16. Potyondy, D.O., Cundall, P.A.: A bonded-particle model for rock. Int. J. Rock Mech. Mining Sci. **41**(8), 1329–1364 (2004)
17. Peng, S.Y., Ng, C.W.W., Zheng, G.: The dilatant behaviour of sand–pile interface subjected to loading and stress relief. Acta Geotech. **9**, 425–437 (2014). https://doi.org/10.1007/s11440-013-0216-9
18. Zhou, J., Jian, Q., Zhang, J., et al.: Coupled 3D discrete-continuum numerical modeling of pile penetration in sand. J. Zhejiang Univ. Sci. A. **13**(1), 44–55 (2012). https://doi.org/10.1631/jzus.a1100172
19. Liu, W.-B., Zhou, J.: Numerical simulation of particle flow code for pile under uplifting load. Chin. J. Geotech. Eng. **26**(4), 516–521 (2004)
20. Lobo-Guerrero, S., Vallejo, L.E.: Influence of pile shape and pile interaction on the crushable behavior of granular materials around driven piles: DEM analyses. Granular Matter **9**(3), 241–250 (2007)
21. Jenck, O., Dias, D., Kastner, R.: Discrete element modelling of a granular platform supported by piles in soft soil—validation on a small scale model test and comparison to a numerical analysis in a continuum. Comput. Geotech. **36**, 917–927 (2009)

22. Lai, H.-J., Zheng, J.-J., Zhang, J., Zhang, R.-J., Cui, L.: DEM analysis of "soil"-arching within geogrid-reinforced and unreinforced pile-supported embankments. Comput. Geotech. **61**, 13–23 (2014)
23. Ng, T.-T., Meyers, R.: Side resistance of drilled shafts in granular soils investigated by DEM. Comput. Geotech. **68**, 161–168 (2015)
24. Villard, P., Chevalier, B., Le Hello, B., Combe, G.: Coupling between finite and discrete element methods for the modelling of earth structures reinforced by geosynthetic. Comput. Geotech. **36**(5), 709–717 (2009)
25. Hamdi, P., Stead, D., Elmo, D.: Damage characterization during laboratory strength testing: A 3D-finite-discrete element approach. Comput. Geotech. **60**, 33–46 (2014)
26. Indraratna, B., Ngo, N.T., Rujikiatkamjorn, C., Sloan, S.W.: Coupled discrete element–finite difference method for analysing the load-deformation behaviour of a single stone column in soft soil. Comput. Geotech. **63**, 267–278 (2015)

Investigation and Numerical Simulation Analyses of the Landslides in Terrace Formation

Sung-Chi Hsu(✉), Ming-Hung Liu, Tai-Seong Quah, and Yishuo Huang

Department of Construction Engineering, Chaoyang University of Technology, Taichung, Taiwan
{schsu,ce,yishuo}@cyut.edu.tw, tai_seong@hotmail.com

Abstract. In situ exploration and sampling are carried out in Sin-San Village in Nantou County after the typhoon. Large direct shear tests in the laboratory are performed to measure the shear strength of the material. The soil water characteristic curve of the soil was derived from the pressure plate tests. The cross-section profiles of the failed slope before and after the typhoon at this area are obtained using digital elevation model (DEM) and 3D terrestrial scanner, respectively. The numerical softwares, SEEP/W and SLOPE/W, are used to analyze the slope stability under the similar rainfall by the typhoon. Different ground water level, river level, and friction angle of the soil are considered to compare the values of factor of safety and to reveal the failure processes of the slope. The possible failure of the slope may be due to scouring of the toe of the slope by the surge of rising river level and/or mud flow based on the investigated profiles and the numerical simulations. The estimated length of scouring was about 20 m long. The possible mechanism of failure for this length of scouring could be due to multiple retrogressive failures after toe scouring based on the numerical modeling.

1 Introduction

In 2009, the moderate Typhoon Morakot invaded Taiwan brought more than 2500 mm of torrential rain in central and southern Taiwan. It was measured as 3059.5 mm in Alishan rainfall station which has been closed to the average rainfall (3684 mm) per year in Alishan. This astonishing rainfall has exceeded the hundred year rainstorm frequency. This was the deadliest typhoon that ever struck Taiwan caused serious disasters in southern and central Taiwan, leaving 461 people dead and 192 others missing. The storm also produced huge amount of rainfall, peaking at 1623, 2361, and 2748 mm in 24, 48, and 72 h, respectively. Some mountain villages in southern Taiwan were covered by the soils from the landslide and debris flow without any warning. Thus, it is very necessary to study and understand the causes of failure and failure mechanisms under extreme weather conditions on the young deposits in these mountain areas.

© Springer International Publishing AG, part of Springer Nature 2019
A. Sevi et al. (eds.), *Enhancements in Applied Geomechanics, Mining, and Excavation Simulation and Analysis*, Sustainable Civil Infrastructures,
https://doi.org/10.1007/978-3-319-95645-9_11

The infiltration of heavy rainfall made the factor of safety of slopes to decrease and caused landslides initiated from the upper and/or downward slopes. The landslides may then cause and trigger mudslides or debris flow. The mud flow or debris flow may squeeze the main river channel and scour river banks and/or down slopes of the roads. Therefore, a lot of landslides had occurred during this typhoon along the road of Tai-21 in central Taiwan. The focus of this research is to study the causes and failure mechanism of slope failures of high terrace deposits in central Taiwan under hydro-geology and extreme weather conditions.

Slope failures due to rainfall conditions have been studied by many researchers and the impacts of unsaturated and saturated soil behavior on the slope stability are also considered in many papers. The objective of this paper is to study and understand the failure processes and mechanisms of the slope failure of high terrace deposits at Shin-Shan Village in Nantou County during Typhoon Morakot. There are thirty-six tributaries that classified as potential debris flow torrents inside this watershed. Thus, a lot of failures and disasters had occurred at the roads and slopes within the Chen-Yu-Lan River watershed. In situ site investigations and sampling have been performed at these areas. Numerical modeling using softwares, SEEP/W and SLOPE/W, developed by Geo-Studio are chosen to simulate the behavior of the slope under both extreme weather and erosion conditions.

2 Disaster Area

The location of the study area is in Nantou County, which situates at the center of Taiwan, and mountain areas cover most of the areas in this county, as shown in Fig. 1. Thus, a lot of disasters, including flooding, slope failures, debris flow, and road and bridge failures, have occurred in this county, especially inside Chen-Yu-Lan Watershed, during Typhoon Morakot. Investigation had been accomplished after the typhoon to document the location and extend of various types of failures. Thus, subsequent researches and analyses could be performed to reveal the causes of disasters. Several major types of disasters, i.e., landslide, debris flow, slope failure due to scouring, have been found during the in situ investigation in Nantou County. Most of the disasters were location beside or close to the Chen-Yu-Lan River.

There are 39 slope related failures occurred in this typhoon, and 24 failures are from upslope and the others are from downslope. This paper will focus on the slope failures at the Shin-Shan Village which is located just beside Chen-Yu-Lan River. The aerial photo taken before the typhoon and location of the village and watershed are show in Fig. 1. The area within the red dashed line was eroded away during the typhoon. There are two potential debris flow torrents, DF169 and DF170, inside this village, and they all flow into Chen-Yu-Lan River. The extreme weather condition induces the quick rise of river level and may cause the flow to turn and erode banks beside the river. The downslope of this area collapsed and fallen down to the river and caused part of the base of road and houses on this terrace to be empty, shown in Fig. 2. Twenty three houses above the slope fell down to the slope into the river because of the slope failure, as shown in Fig. 2. Both sides of the river banks have been eroded away and the river channel was widen during the typhoon. Kriging interpolation method is

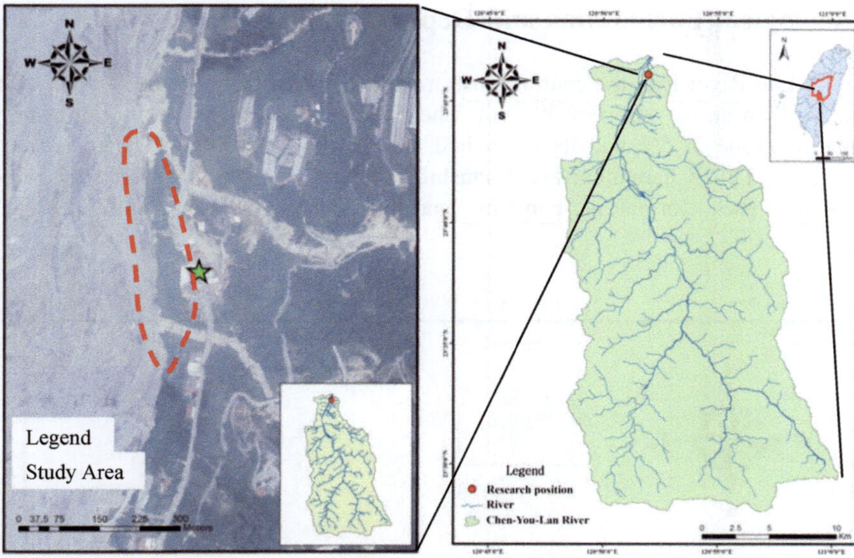

Fig. 1. Locations of study area and Chen-Yu-Lan River Watershed (Aerial Photo Taken in 2005)

used to obtain the hourly rainfall during the typhoon based on thirteen rainfall stations inside Chen-Yu-Lan River Watershed because there is no rainfall station near the village.

Fig. 2. Downslope of Shin-Shan Village collapsed and near fallen houses beside the river

3 Geological Formations and Deposits

Chen-Yu-Lan River is at the central Taiwan and originates from the north peak of Yu Mountain with an elevation of 3910 m. The river length is 42.4 km with an average declination slope of 5% and its watershed area is about 45,000 ha. There are three major faults, Dili, Chenyulan, and Shanshihchia Faults across this watershed which causes the rock formation contains many fractures and discontinuities. The

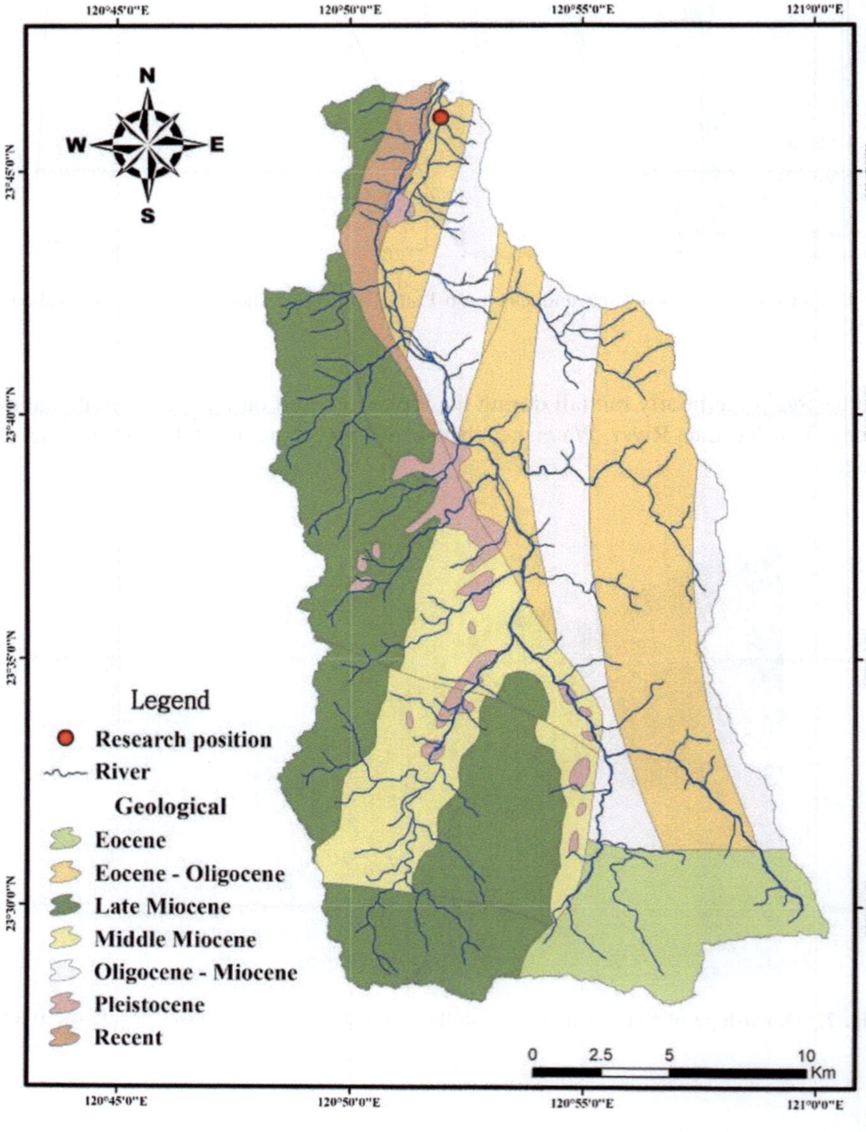

Fig. 3. Geological formations of Chen-Yu-Lan River watershed (After Hsu et al. 2011)

Chenyulanchi fault, which almost parallels Chen-Yu-Lan River, separates the Miocene sedimentary rocks of the Western Foothills from the east side of the Oligocene metamorphic rocks of the Shuehshan Range (Fig. 3). Differential uplifting along this fault has generated great topographic relief and abundant fractures that resulted in frequent landslides and debris flows even prior to the Chi-Chi earthquake (Lin et al. 2002). Slates and meta-sandstones are the dominant lithologies in the metamorphic terrane. Based on the relative amount of slate and meta-sandstone, the metamorphic strata are divided into four formations: in ascending order, the Shihpachuangchi, the Tachien Meta-Sandstone, the Paileng Meta-Sandstone, and the Shuichangliu. In the sedimentary terrane, sandstone and shale predominate and include the Nankang, Nanchuang, and Kueichulin Formations (Lin et al. 2002). The slope failure at Sin-San Village is located in the recent deposit and can be classified as terrace deposit. This terrace deposit could be formed by previous flood, debris flow, and/or sliding materials.

4 Characteristics of Terrace Deposit

4.1 Index Properties

The grain-size distribution and unit weight of the soil in Shin-Shan Village are obtained from the trench excavation (50 × 50 × 50 cm). The weight and volume of the soil are measured, and sieve analyses are performed. The dry unit weight is 20.4 kN/m^3 and the biggest grain size is around 30 cm. The soil is classified as poorly graded gravel, GP. The percentages of gravel and sand are about 68 and 32%, respectively. Therefore, the terrace deposit at Sin-San Village is considered as gravel formation. Gravel formations are common in western Taiwan, especially in central Taiwan and for these high terrace deposits.

4.2 Characteristic Curve and Conductivity

In order to model the impacts of rainfall infiltration on slope stability of unsaturated soils, the soil water characteristics curve is required for numerical simulation. Pressure plate test is used to obtain the soil water characteristics curve, shown in Fig. 4, for the soil for modeling of rainfall infiltration. The van Genuchten (VG) model is used to fit the soil water characteristic curve of the soil, as shown in Fig. 4. The equation obtained by Wu et al. (2012) is used to predict the hydraulic conductivity of the soil. They used single-ring infiltrometer to estimate the hydraulic conductivity using percentage of gravel content, porosity, and effective particle size D_{10}. The estimated conductivity for this terrace deposit is about 0.015 cm/s.

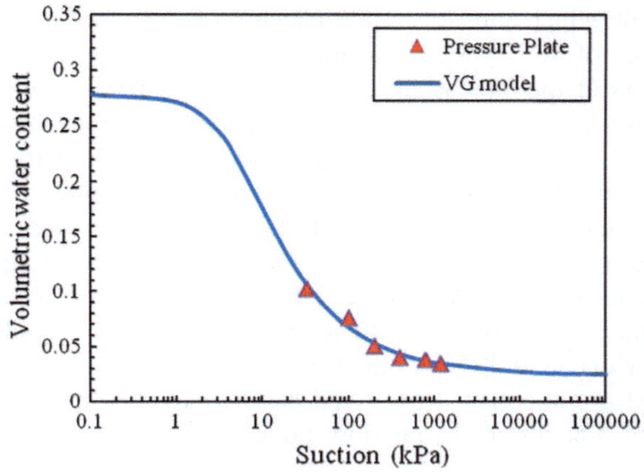

Fig. 4. Volumetric water content versus matric suction and fitted VG model

4.3 Measurement of Failed Profile

A Terrestrial Scanner (VZ-1000) is used to measure the slope profile after the failure caused by the Typhoon Morakot. The cross-section profiles of the failed slope before the typhoon at this area was obtained using digital elevation model (DEM). The results of the profiles by 3D terrestrial scanner are shown in Fig. 5. About 20 m of the slope near the toe area was eroded away during the typhoon by comparing the two slope profiles.

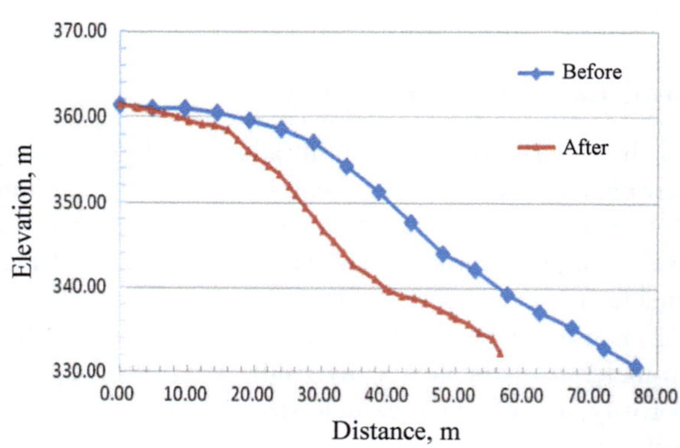

Fig. 5. Cross sectional profiles before and after the failure

5 Shear Strength of Terrace Deposit

The maximum grain size of this terrace deposit is 30 cm based on the in situ sampling. Thus, the equal-weighted method is used to reduce the maximum grain size of the tested samples to 10 cm and adjust the grain-size distribution accordingly in order to perform large direct shear test. Modified and standard compaction tests are made to obtain the compaction energy similar to the in situ soil condition. Then, a large direct shear tests (50 × 50 cm) are used to acquire the shear strength of the terrace deposit under saturated and unsaturated conditions. The shear strength of the sample with 5% water content is larger than the saturated sample. According to the in situ large direct shear test, 1.5 × 1.5 m (Chu et al. 1989) performed in central Taichung, the peak and residual friction angles were 54.3° and 44.9°, and apparent cohesions were 14.7 and 0 kN/m^2, respectively. The tested results of samples with 5% water content are used to compare and combine with the in situ large direct shear test and plotted in Fig. 6. The peak and residual friction angles are about 54° and 46°, respectively. The peak apparent cohesion is about 25 kPa. Considering the shear strength may decrease as water content increases during rainfall. Therefore, the friction angle and apparent cohesion of the terrace deposit used for later numerical modeling are chosen as 46° and 15°, respectively.

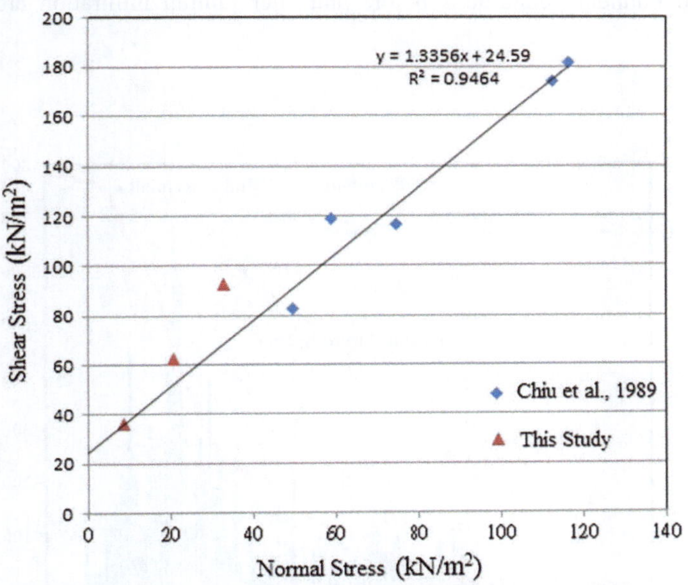

Fig. 6. Results of direct shear tests from this study and Chiu et al. (1989)

Table 1. Parameters used for SEEP/W and SLOPE/W

Head water level	Tail water level	Saturated water content	Hydraulic conductivity	Unit weight	Cohesion	Friction angle
335 m	330 m	0.29	0.015 cm/s	19.7 kN/m^3	15 kPa	46°

6 Numerical Modeling

6.1 Parameters Used for Modeling

In order to model the impacts of rainfall infiltration on the slope stability and possible failure processes of high terrace deposits, numerical software is selected to simulate slopes under extreme weather condition. Therefore, numerical modeling using software, SEEP/W and SLOPE/W, developed by Geo-Studio (Geo-Slope 2013), are chosen to model infiltration and slope stability, respectively. The VG model is used for the material during analyses. The shear strength, water levels and soil properties used for SEEP/W and SLOPE/W are listed in Table 1. The rainfall duration and intensity during Typhoon Morakot is used for SEEP/W, as shown in Fig. 7. The schematic profiles and boundary conditions before and after rainfall infiltration are shown in Fig. 8.

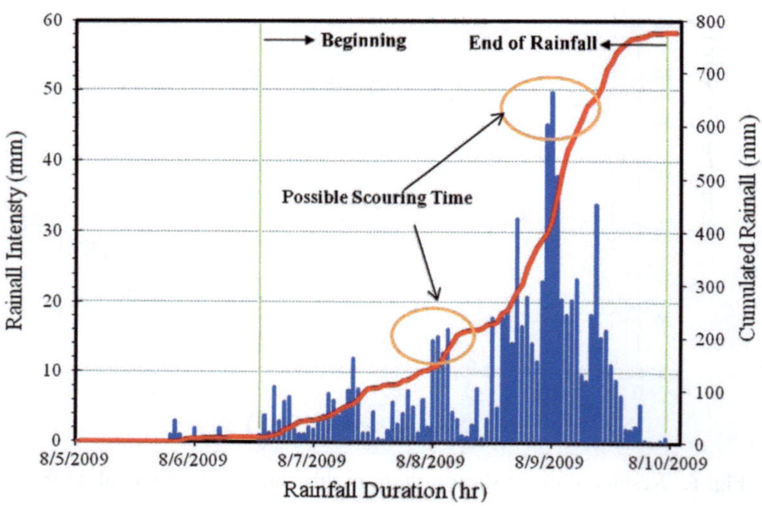

Fig. 7. Rainfall duration and intensity during Typhoon Morakot

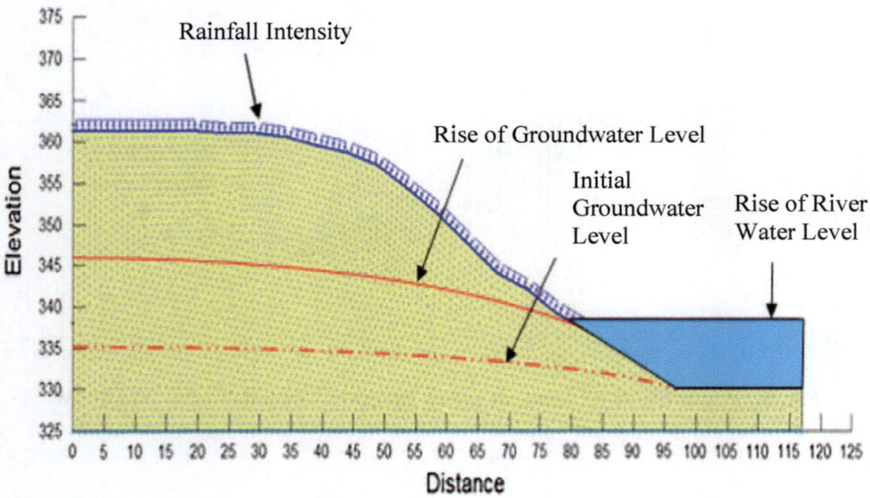

Fig. 8. The schematic profiles and boundary conditions before and after rainfall infiltration

7 Results and Discussion

The slope is modeled using SEEP/W under the rainfall intensity in Fig. 7, and then uses SLOPE/W to calculate the factor of safety at each time. The factor of safety for the slope before rainfall is 2.61. The lowest factor of safety for the slope during rainfall without considering change of river level is 1.23 after 60 h of raining, i.e. the slope is still stable. According to the observation from the residents in that village, the river level goes up very quick and started scouring the toe of the slope during the typhoon. Thus, the river level is also assumed to go up as the groundwater level rose due to infiltration. The tail groundwater level rises from 330 to 343 m after 60 h of rainfall, and the head groundwater level rises from 335 to 350 m. The lowest factor of safety for the slope considering rising of river water level is down to 1.187, as shown in Fig. 9, after 60 h of rain. The lowest factor of safety for the slope will be down to 0.915 if the analysis considering rise of river water level and scouring effect at the toe pf the slope. Thus, retrogressive failure will be initiated since the factor of safety is less than 1. At least three subsequent sliding failures could occur after the first slide. Figure 10 shows slope profile and the factor of safety after the toe erosion and the first slide. The factor of safety is 0.937. The slope is still unstable and will slide again. The obtained slope profile after four sliding failure is similar to the actual failed profile after the typhoon. Therefore, the major cause of the slope failure could be due to scouring and erosion of the river on the toe of the slope.

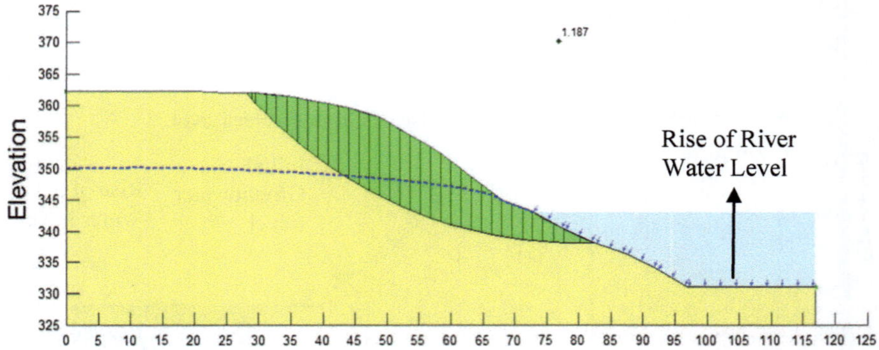

Fig. 9. Profile of lowest factor of safety after 60 h of rainfall using SLOPE/W

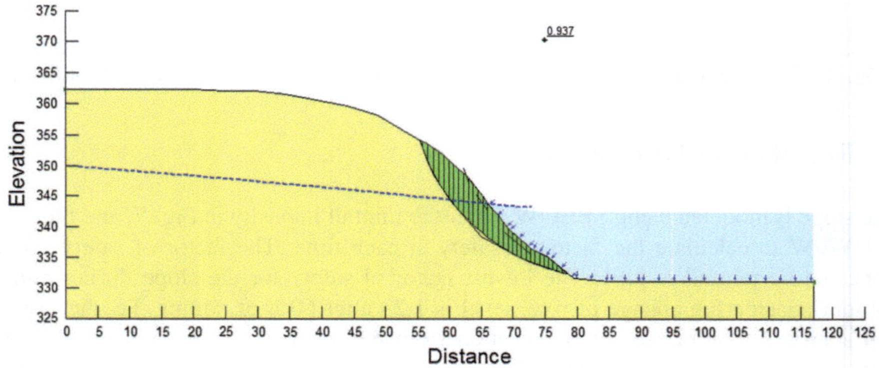

Fig. 10. Profile of lowest factor of safety after toe erosion and first slide

8 Conclusions

Failure of high terrace deposit during Typhoon Morakot beside Chen-Yu-Lan River, Nantou County, Taiwan is chosen for this study. The following conclusions can be drawn based on the study and numerical modeling:

1. Part of the village is eroded away based on the aerial photos taken before and after the typhoon.
2. Soil water characteristic curve is obtained from the pressure plate tests and fitted with VG model.
3. The slope could remain stable if the river water level had not gone up and the toe of the slope was not eroded away during Typhoon Morakot.
4. The factor of safety will decrease if the river water level goes up.

5. The slope will become unstable if the toe of the slope is eroded away due to the river flow.
6. Retrogressive failure will occur if the river water level goes up and the toe of the slope is eroded away.

Acknowledgements. The authors would like to thank National Science Foundation in Taiwan for financially supporting this research. Also thanks to Nantou Branch of Directorate General of Highways of Taiwan for providing the valuable information and photos.

References

Chu, B.-L., Hwang, C.L., Cheng, S.Y., Pan, G.M.: Study of in situ direct shear tests in gravel formation in Taiwan. In: The Third Geotechnical Conference, Taiwan, pp. 695–706 (1989)

Geo-Slope: User's Manual for SEEP/W and SLOPE/W: Version 4. Geo-Slope International, Canada (2013)

Hsu, S.-C., Lai, B.J., Hsu, W., Lai, J.: Modeling of gravel properties and anchors using discrete element method. Adv. Mater. Res. **189–193**, 1726–1731 (2011)

Lin, P.-S., Lin, J.-Y., Hung, J.-C., Yang, M.-D.: Assessing debris-flow hazard in a watershed in Taiwan. Eng. Geol. **66**, 295–313 (2002)

Wu, C.-C., Wang, C.-H., Chan, Y.-T.: Infiltration behavior of gravel-rich soil. J. Chin. Soil Water Conserv. **43**(1), 65–74 (2012)

Performance of Three Atmospheric Density Models on Precise Orbit Determination for Haiyang-2A Satellite Using DORIS Data

Qiaoli Kong[1,2,3(✉)], Jinyun Guo[1,2], Litao Han[1], and Yi Shen[1]

[1] College of Geomatics, Shandong University of Science and Technology, Qingdao, China
kqlabc3334@hotmail.com, jinyunguo1@126.com

[2] State Key Laboratory of Mining Disaster Prevention and Control Co-Founded by Shandong Province and the Ministry of Science and Technology, Shandong University of Science and Technology, Qingdao, China

[3] Engineering Laboratory of Spatial Information Technology of Highway Geological Disaster Early Warning in Hunan Province, Changsha University of Science & Technology, Changsha, China

Abstract. DORIS has become a matured space geodetic technique after more than ten years of development. This geodetic technique is mainly applied to determine the orbit for the low Earth orbit (LEO) satellite. There are a number of non-conservative forces acting on Haiyang-2A (HY-2A) satellite with altitude of about 970 km, in which the atmospheric drag is the most dominant and uncertainty in the precise orbit determination (POD) with the dynamic method. In order to achieve POD for HY-2A, MSIS-86, Jacchia 1971 and DTM87 models were evaluated in this study. The precise orbits of HY-2A from DORIS data were compared with the precise orbit ephemeris computed by the Centre National d'Etudes Spatiales (CNES). Tests demonstrated that the relative optimal atmospheric density model was the empirical MSIS-86 model for HY-2A satellite with the corresponding drag coefficient of 2.0. The RMS of orbit difference between the derived orbits using MSIS-86 and the CNES orbits was 0.842 cm in the radial direction, and 3.899 cm in three dimensions (3D). This study will provide valuable references for the LEO satellites with similar altitude and surface, especially for the other HY-2 series satellite of China.

1 Introduction

Haiyang-2A Satellite (HY-2A) was launched on August 16, 2011. It is the first satellite designed to monitor the global marine dynamic environment with microwave sensors and detect ocean surface wind field, height and temperature in China. It is also the first mission of China's Earth observation satellite HY-2 series (including HY-2A, HY-2B, HY-2C, and HY-2D) and China's first radar altimetry satellite. Precision orbit determination (POD) is an essential objective for HY-2A to provide fundamental references for the analysis and application of the on-board equipment. In order to achieve the essential goal, HY-2A satellite carries three kinds of instrument for POD activities, that is, a Doppler Orbit Determination and Radio positioning Integrated by Satellite

© Springer International Publishing AG, part of Springer Nature 2019
A. Sevi et al. (eds.), *Enhancements in Applied Geomechanics, Mining, and Excavation Simulation and Analysis*, Sustainable Civil Infrastructures,
https://doi.org/10.1007/978-3-319-95645-9_12

(DORIS) receiver, a dual frequency Global Positioning System (GPS) receiver and a Satellite Laser Ranging (SLR) retro-reflector array (Guo et al. 2013a; Kong et al. 2014).

DORIS is an excellent geodetic technique supporting precise orbit determination (POD) of the low Earth orbit satellite (LEO) and geodesy (Zelensky et al. 2010). The current DGXX generation DORIS receiver onboard HY-2A and Jason-2 can simultaneously track 7 DORIS beacons with 2 × 7 channels (Mercier et al. 2010), and range-rate and phase measurements on both channels with RINEX format can be provided at the same epoch. The evenly distributed beacons and the capability of tracking 7 DORIS beacons can improve the geometry strength and increase the observation quantity significantly from the former generation DORIS receiver.

There are a number of non-conservative forces acting on a LEO satellite, in which the atmospheric drag is the most dominant and also contains the most uncertainty. The atmospheric drag is directly proportional to atmospheric density. There are many factors affecting the variability of atmospheric density, and some of these factors have not been well modeled, such as atmospheric heating and the solar and geomagnetic activity levels. The Earth's upper atmospheric density changes with the solar activity, time, season, latitude and geomagnetic index, and the mechanisms are too complex to be accurately modeled. Atmospheric drag is one dominant constrain on the accurate of orbit determination. It is very important to make a full research on the atmospheric density model and atmospheric drag coefficient to achieve POD for the LEO satellite using DORIS data. At present, significant researches have taken place to determine the proper modeling for atmospheric density. Fattig (2011) demonstrated that the RMS differs very little when the Jacchia family of atmospheric density models, namely CIRA 1972, Jacchia (1971), and Jacchia-Robert (Robert 1971) were used to determine the precise orbits of GRACE-A and GRACE-B, and the test showed that the Jacchia 1971 was the optimal model in the Jacchia family. Apart from the well-known and frequently applied Jacchia model, a variety of other density models of the upper atmosphere exits, such as MSIS (Mass Spectrometer and Incoherent Scatter) and DTM (French Density Temperature Model) families. Li et al. (2010) found that the impact of MSIS-86 (Hedin 1987) atmosphere density model on the accuracy of the orbit varied largely with different altitude, and the orbital accuracy in the radial direction is about 1.5 cm, and the one in along track direction is about 3.0 cm using DORIS technique. Mance (2010) precisely determined orbits of GEOSAT Follow-On (GFO), Starlette, Stella, and Geo-Forschungs-Zentrum-1 (GFZ-1) satellites using MSIS-86 atmospheric density model. Lechtenberg (2010) had tested the performance of different atmospheric density models (such as Jacchia, MSIS and DTM families) using GPS technique for the POD of CHAMP and using SLR technique for the POD of GRACE, and found that the Jacchia based models outperformed the other two density model families. Lechtenberg (2015) estimated atmospheric densities using SLR data and made a full research on the density correction factors and standard deviations for the ANDE Castor satellite.

In order to realize precise orbit determination for the HY-2A satellite, several researches have presented their strategies using different geodetic techniques and different atmospheric density models. Zhao et al. (2013) and Wang et al. (2014) have used the DTM94 (Berger et al. 1998) atmospheric model and have achieved the radial accuracy of better than 3 and 4 cm using SLR geodetic technique, respectively. Guo et al. (2013b) and Lin et al. (2014) have used the same model and have got 1–2 cm and

3 cm radial accuracy using GPS technique, respectively. Kong et al. (2014) have applied DTM87 model (Barlie et al. 1978) and about 1.5 cm accuracy has been gained in radial direction using DORIS technique. Gao et al. (2015) have used MSIS-86 atmospheric density model and accuracy about 1.1 cm has been obtained in radial direction with DORIS data. Both DTM and MSIS atmospheric density family models were applied based on SLR, GPS and DORIS data for HY-2A, respectively.

Many studies and even some international standards have promoted one model over another, while it is necessary to select the best model for a particular mission. The goal of this research is to invest the performance of MSIS-86, Jacchia 1971 and DTM87 density models, and thus to find a relatively accurate coefficient and density model to enhance atmospheric drag calculations for the POD of HY-2A satellite.

2 Atmospheric Drag Modeling

The observed atmospheric drag forces act in the opposite direction of the satellite velocity. Accurately modeling atmospheric forces is difficult because of the difficulty of availability of the physical properties, complexity of the interaction of neutral gas with different spacecraft surfaces, and variation of attitude of spacecraft. Usually satellite surface is divided into a number of limited basis units to calculate the atmospheric drag on per unit area. The widely accepted equation for atmospheric drag is

$$\Delta \ddot{r}_d = -\frac{1}{2} C_D \frac{A}{m_s} \rho v_r^2 \hat{v}_r \tag{1}$$

where the negative sign indicates that the acceleration is anti-parallel to the unit relative velocity vector ($\hat{v}_r$), which provides the direction of the acceleration. C_D is introduced to model the actual momentum transfer dependent upon the interaction between the satellite and the atmosphere. Typically, the nominal value of C_D ranges from 2.0 to 2.5 (Montenbruck and Gill 2000). A is the cross sectional area of the orbiting body and is normal to the velocity vector. m_s is the mass of the orbiting body, and v_r is the velocity of the satellite.

3 POD Strategy for HY-2A Using DORIS Technique

Considering the amount of tracked DORIS data and strength of geometric structure of beacons, we should apply the dynamic method to determine the orbit for HY-2A satellite. Six prior independent parameters at the initial epoch are needed for the orbit integration for the dynamic method. These six parameters can be the three-dimensional position and velocity vectors or six orbital elements. HY-2A is disturbed by the conservative forces and the non-conservative forces as this mission travels around the Earth. The conservative forces include the Earth's gravitation, n-body perturbation, oceanic and solid earth tides, et al., and the non-conservative forces include the solar and earth radiations, atmospheric drag and relativistic effect, and other small perturbing factors. The multi-step COWELL II numerical integration was used during the process of orbit determination for HY-2A (Balmino et al. 1990), and the step size for orbit integration is 10 s and the output interval of orbit is set to 60 s. During the orbit determination, the empirical force model was applied to compensate the other orbit dynamic perturbations which cannot be modeled and estimated (Kang et al. 2006; Liu 2013). In the process of parameter estimation using the DORIS Doppler tracking data, all parameters are solved together, such as the coordinate, velocity, solar radiation coefficients and empirical acceleration coefficients. The initial state vectors and empirical acceleration coefficients in the cross track are estimated, and the coefficients of solar radiation are estimated every 6 h. The least squares estimation method is applied to solve the unknown parameters. The cutoff elevation angle is set to 10°. Due to TAI time system used in DORIS system, the constant offset (leap second) of 35 s is introduced to be in consistent with UTC time. Table 1 summarizes the applied dynamic models.

Table 1. Dynamic models for precise orbit determination of HY-2A

Items	Description
Coordinates of DORIS beacon stations (Willis et al. 2012, 2013) and DORIS Doppler measurements	http://www.ipgp.fr/~willis/DPOD2008/, ftp://cddis.gsfc.nasa.gov/pub/doris/data/h2a
Earth gravity model	EGM2008 (Pavlis et al. 2012), 80 × 80
N-Body	JPL DE403 (Standish 1998)
Solid earth tides	IERS2010 (Petit and Luzum 2010)
Ocean tides and ocean tides loading	FES2004 (Lyard et al. 2006)
Relativistic effect	IERS2003 (McCarthy and Petit 2003)
Solar radiation pressure	Box-Wing (Rim 1992)
Earth albedo radiation	Knocke–Ries–Tapley (Knocke et al. 1988)
Tropospherical model	Hopfied (Goad and Goodman 1974)
Atmospheric drag	MSIS-86 (Hedin 1987), Jacchia 1971 (Jacchia 1971), DTM87 (Barlier et al. 1978)

In this paper, we applied DORIS range rate data to compute the precise orbit of HY-2A. During the 3-day arc from 8 Sep., 2012 to 10 Sep., 2012, HY-2A was tracked by 48 DORIS stations. Figure 1 displays the geographical location of DORIS beacon stations and the 3-day trajectory of HY-2A.

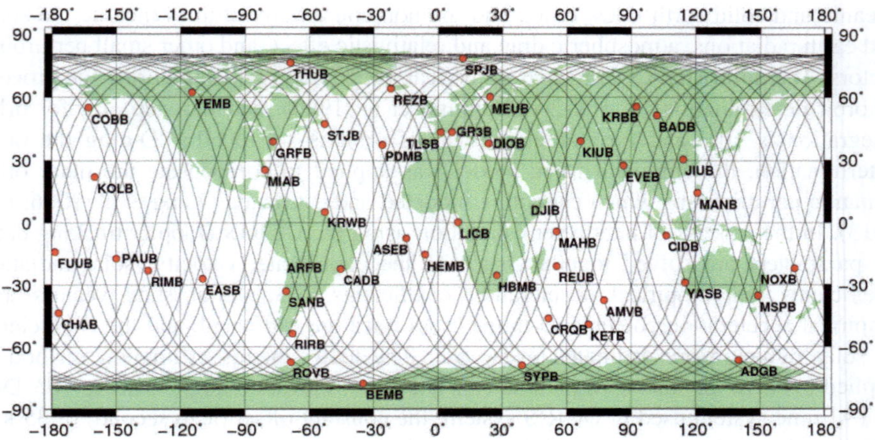

Fig. 1. Distribution of DORIS tracking stations on the global and the trajectory of HY-2A from September 8 to 10, 2012

4 Performance of Different Atmospheric Models and Analyses

The coefficient of drag C_d is ultimately difficult to define for the complex shape of HY-2A satellite. The determination of the C_d is related to several parameters such as the accommodation coefficient, momentum accommodation coefficient, the temperature of the atmosphere, the temperature of the satellites surface, altitude, the velocity of the satellite, solar minimum and solar maximum, as well as other parameters (Mehta et al. 2014; Pilinski et al. 2011). Drag coefficients for satellites in the upper atmosphere are typically approximated from 2.0 to 2.5. The empirical drag coefficients are 2.3 and 2.5, while the coefficient also has the close relation to the altitude of the satellite. The popular density models, such as Jacchia 1971, MSIS-86 and DTM87 atmospheric density model, were applied to compute the orbit for HY-2A. In order to search for the optimal coefficients for this mission, we have estimated the orbits using MSIS-86, Jacchia 1971 and DTM87 model with the coefficients of 2.0, 2.1, 2.2, 2.3, 2.4 and 2.5 for HY-2A satellite with 970 km altitude, respectively.

The statistics of RMS differences between the CNES orbit and the orbit derived from MSIS-86 with different atmospheric drag coefficients are showed in Table 2.

Table 2 indicates that the smallest RMS value is 0.842 cm in radial direction computed using coefficients of 2.0, 2.2, 2.3, respectively, and the biggest RMS value is computed using MSIS-86 with coefficients 2.4 and 2.5. The smallest and the biggest RMS values in along track direction are 2.682 and 2.893 cm achieved from MSIS-86

Table 2. Statistics of orbit difference between the CNES orbits and the derived orbits using MSIS-86 model with different coefficients (cm)

Direction	Radial		Along track		Cross track		3D
	Mean	RMS	Mean	RMS	STD	RMS	RMS
2.0	−0.118	0.842	0.690	2.682	2.701	2.702	3.899
2.1	−0.118	0.843	0.692	2.686	2.702	2.703	3.903
2.2	−0.118	0.842	0.695	2.688	2.704	2.705	3.905
2.3	−0.120	0.842	1.050	2.884	2.694	2.696	4.037
2.4	−0.120	0.843	1.049	2.887	2.695	2.697	4.040
2.5	−0.119	0.843	1.053	2.893	2.697	2.699	4.045

with coefficient 2.0 and 2.5 respectively. The smallest RMS in 3D is 3.899 cm achieved using this density model with coefficient 2.0. From Table 2, we can notice that the relative optimal accuracy of the orbit is derived using the empirical atmospheric density model MIS-86 with drag coefficient 2.0, which has the closest agreement with the precise CNES orbit, and the differences between both orbits are listed in Fig. 2. Therefore, 2.0 is the relative optimal coefficient for MSIS-86 model for HY-2A satellite. At the same time, we can know that there is little change with the RMS in radial direction and cross track direction from the coefficient 2.0 to 2.5, and there are large changes in along track direction. The main reasons are that the atmospheric drag acts on the satellite conversely along the track direction, and it is in proportion to the atmospheric density, and the change of drag coefficient will bring change of atmospheric drag.

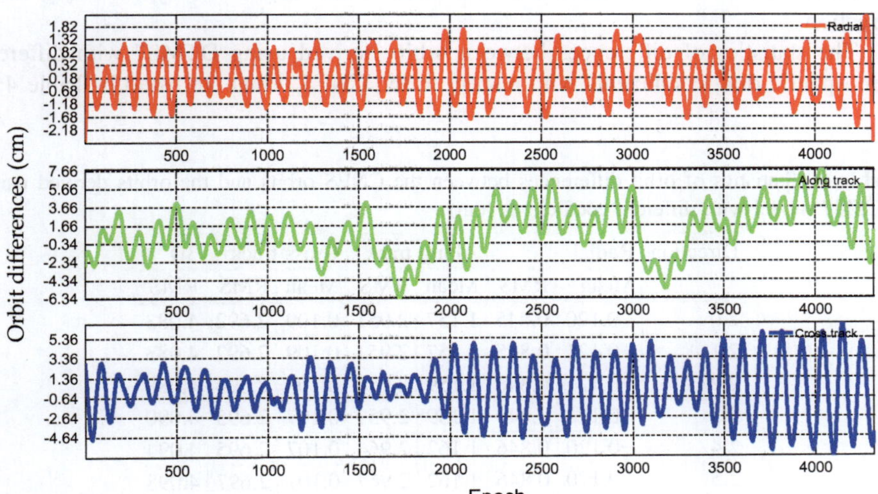

Fig. 2. The orbit difference between the derived orbit using MSIS-86 model with coefficient 2.0 and the CNES orbit

The statistics of differences between the CNES orbits and the orbits derived from Jacchia 1971 with different atmospheric drag coefficients are listed in Table 3.

Table 3. Statistics of orbit difference between the CNES orbits and the derived ones using Jacchia 1971 model with different coefficients (cm)

Direction	Radial		Along track		Cross track		3D
	Mean	RMS	Mean	RMS	Mean	RMS	RMS
2.0	−0.120	0.842	1.146	2.933	0.102	2.693	4.070
2.1	−0.121	0.842	1.148	2.936	0.102	2.694	4.072
2.2	−0.121	0.842	1.149	2.939	0.102	2.695	4.075
2.3	−0.121	0.842	1.150	2.942	0.100	2.696	4.079
2.4	−0.120	0.843	1.049	2.887	0.102	2.697	4.040
2.5	−0.120	0.842	1.048	2.880	0.102	2.700	4.037

Table 3 indicates that the smallest RMS value is 0.842 cm in radial direction computed using coefficients of 2.0, 2.1, 2.2, 2.3 and 2.5, respectively, and the biggest RMS is 0.843 cm derived from the Jacchia 1971 with coefficient 2.4. The smallest RMS value in along track direction is 2.880 cm derived from this model with coefficient 2.5, and the biggest one is 2.942 cm achieved with coefficient 2.3. The smallest RMS in 3D is 4.037 cm achieved using this density model with coefficient 2.5. From Table 3, we can see that the relative optimal accuracy of the orbit is derived using the empirical atmospheric density model Jacchia 1971 with drag coefficient 2.5. Therefore, 2.5 is the relative optimal coefficient for Jacchia 1971 model for HY-2A satellite. At the same time, we can know that there are the smallest changes with the RMS in radial direction and cross track direction, and there is the largest change in along track direction.

The statistics of differences between orbits derived using DTM87 with different atmospheric drag coefficients with respect to the CNES orbits are listed in Table 4.

Table 4. Statistics of orbit differences between the CNES orbits and the orbits derived using DTM87 model with different coefficients (cm)

Direction	Radial		Along track		Cross track		3D
	Mean	RMS	Mean	RMS	Mean	RMS	RMS
2.0	−0.120	0.845	1.157	2.951	0.109	2.692	4.083
2.1	−0.121	0.845	1.157	2.951	0.109	2.692	4.083
2.2	−0.121	0.846	1.159	2.956	0.108	2.694	4.088
2.3	−0.121	0.845	1.159	2.959	0.107	2.695	4.090
2.4	−0.120	0.846	1.162	2.964	0.107	2.695	4.094
2.5	−0.120	0.846	1.162	2.967	0.107	2.697	4.098

Table 4 shows that the smallest RMS value is 0.845 cm in radial direction computed using coefficients 2.0, 2.1 and 2.3, respectively, and the biggest one is 0.846 cm derived from DTM87 with coefficients 2.2, 2.4 and 2.5, respectively. The smallest RMS value in along track direction is 2.951 cm derived with Jacchia 1971 with coefficient 2.0 and 2.2, respectively, and the biggest one is 2.967 cm achieved using this model with coefficient 2.5. The smallest RMS in 3D is 4.083 cm computed using this density model with coefficients 2.0 and 2.1, respectively. From Table 4, we can see that the relative optimal accuracy of the orbit is derived using the empirical atmospheric density model DTM87 with drag coefficient 2.0. At the same time, we can know that the RMS in radial direction and cross track direction vary little between the results derived using different coefficients, and vary largely in along track direction.

5 Conclusions

The work was performed to determine the relative optimal drag coefficients and drag density model for the HY-2A satellite by using DORIS data. In order to obtain the optimal solution, we compared the performance of different coefficients of atmospheric drag using different atmospheric density models. We computed orbits using MSIS-86, Jacchia 1971 and DTM87 atmospheric models applying DORIS technique, respectively, and analyzed the orbit difference RMS with respect to CNES orbits. The results indicate that the empirical atmospheric density model MSIS-86 performed significantly better than the other two density models. Among the coefficients of 2.0, 2.1, 2.2, 2.3, 2.4 and 2.5, we found that the relative optimal atmospheric coefficient is 2.0 for MSIS-86 density model, and 2.5 for Jacchia 1971 density model. The smallest RMS difference with respect to the CNES orbit is 0.842 cm in radial direction, and the biggest RMS is 2.702 cm in cross track direction computed using MSIS-86 with coefficient 2.0.

There are mainly three possible reasons for the RMS of orbit difference better than 1 cm derived from DORIS data with respect to the CNES orbit. Firstly, the DORIS beacon stations are distributed evenly; secondly, there are large amount of measurements of DORIS; finally there are more tracking stations. From the comparison between different atmospheric models, we can know from Fig. 2 that at the beginning and end of the arc, there are diverse phenomena, and this is mainly because the uncertainties are significantly higher at the beginning of the arc than the remnant part, and the integrator is not stable, and the uncertainties accumulate with the time goes by.

It should be recognized that DORIS, as another space geodetic technique besides GPS, SLR and VLBI, can be used to achieve high precise orbit. In our future work, we will put more research on the improvement of atmospheric density model and investigation of the relationship between atmospheric drag and the latitude, longitude, altitude, the plate face and so on, and realize the POD for the other satellites of HY-2 series.

Acknowledgements. This work was supported by the National Natural Science Foundation of China (Grant No. 41704015, 41774001), A Project of Shandong Province Higher Educational Science and Technology Program (J17KA077), the Public Benefit Scientific Research Project of

China (Grant No. 201412001), International Science and Technology Cooperation Program of China (Grant No. 2009DFB00130), the Basic Science and Technology Research Project of China (Grant No. 2015FY310200), and Open Fund of Engineering Laboratory of Spatial Information Technology of Highway Geological Disaster Early Warning in Hunan Province (Changsha University of Science & Technology) (Grant No. kfj150605).

References

Balmino, G., et al.: Numerical integration techniques revisited. J. Geod. (1990)

Berger, C., et al.: Improvement of the empirical thermospheric model DTM: DTM94—a comparative review of various temporal variations and prospects in space geodesy applications. J. Geod. **72**, 161–178 (1998)

CIRA.: COSPAR international reference atmospheric 1972. In: COSPAR Committee for CIRA. Akademie Verlag, Berlin (1972)

Fattig, E.: Comparison of precision orbit derived density estimates for CHAMP and GRACE satellites. Master's thesis, University of Kansas, Kansas (2011)

Gao, F., et al.: Analysis of HY2A precise orbit determination using DORIS. Adv. Space Res. (2015). https://doi.org/10.1016/j.asr.2014.11.032 (Springer)

Goad, C.C., Goodman, L.: A modified Hopfield tropospheric refraction correction model. In: American Geophysical Union Annual Fall Meeting, San Francisco, USA (1974)

Guo, J., et al.: On precise orbit determination of HY-2 with space geodetic techniques. Acta Geophys. (2013a). https://doi.org/10.2478/s11600-012-0095-8 (Springer)

Guo, J., et al.: Centimeter level orbit determination for HY2A using GPS data (in Chinese). Geomatics Inf. Sci. Wuhan Univ. (2013b)

Hedin, A.E.: A revised thermospheric model based on mass spectrometer and incoherent scatter data MSIS-83. J. Geophys. Res. (1983). https://doi.org/10.1029/ja088ia12p10170 (Springer)

Hedin, A.E.: MSIS-86 thermospheric model. J. Geophys. Res. (1987). https://doi.org/10.1029/JA092iA05p04649

Jacchia, L.G.: Revised static models of the thermosphere and exosphere with empirical temperature profiles. Smithsonian Astrophysical Observatory, Special Report (1971)

Kang, Z., et al.: Precise orbit determination for the GRACE mission using only GPS data. J. Geod. **80**, 322–331 (2006). https://doi.org/10.1007/s00190-006-0073-5

Knocke, P.C., et al.: Earth radiation pressure effects on satellites. In: Proceedings of the AIAA/AAS Astrodynamics Conference, Minneapolis, Minnesota (1988)

Kong, Q., et al.: Precise orbit determination and accuracy analysis of HY-2A satellite using DORIS Doppler data. Acta Geod. Geophys. (2014). https://doi.org/10.1007/s40328-014-0066-4 (Springer)

Kong, Q.: Centimeter level precise orbit determination of HY-2a satellite using DORIS and SLR techniques. Doctoral thesis, Shandong University of Science and Technology, Qingdao (2015)

Lechtenberg, T.F.: Derivation and observability of upper atmospheric density variations utilizing precision orbit ephemerides. Master's thesis, University of Kansas, Kansas (2010)

Lechtenberg, T.F.: Density model corrections derived from orbit data to characterize upper atmospheric density variations. Doctoral thesis, University of Kansas, Kansas (2015)

Li, P.: High accuracy orbit determination with DORIS system. J. Spacecr. TT&C Technol. (2010). 1674-5620(2010)03-0058-07 (Springer)

Lin, M., et al.: Precise orbit determination technology based on dual-frequency GPS solution for HY-2 satellite. Eng. Sci. (2014)

Liu, J.: Satellite orbit determination method with DORIS system. Master's thesis. National University of Defense Technology, Changsha, China (2013)

Locke, T.C.: Further developments in orbit ephemeris derived neutral density. Master's thesis, University of Kansas, Kansas (2012)

Lyard, F., et al.: Modeling the global ocean tides: modern insights from FES2004. Ocean Dynam. (2006). https://doi.org/10.1007/s10236-006-0086-x (Springer)

Mance, S.: An investigation into satellite drag modeling performance. Master's thesis, University of Kansas, Kansas (2010)

McCarthy, D.D., Petit, G.: Equations of motion for an artificial earth satellite. In: IERS Conventions 2003. IERS Technical Note 32, Paris, France (2003)

Mehta, P.M., et al.: Comparing physical drag coefficients computed using different gas-surface interaction models. J. Spacecr. Rockets. (2014). https://doi.org/10.2514/1.a32566 (Springer)

Mercier, F., et al.: Jason-2 DORIS phase measurement processing. Adv. Space Res. (2010). https://doi.org/10.1016/j.asr.2009.12.002 (Springer)

Montenbruck, O., Gill, E.: Satellite Tracking and Observation Models, Satellite Orbits. Springer, Berlin, Heidelberg, German (2000)

Pavlis, N.K., et al.: The development and evaluation of the earth gravitational model 2008 (EGM2008). J. Geophys. Res. (2012). https://doi.org/10.1029/2011jb008916 (Springer)

Petit, G., Luzum, B.: IERS conventions (2010). Bureau International des Poids et Mesures, Sevres, France (2010)

Pilinski, M.D. et al.: Drag coefficients of satellites with concave geometries: comparing models and observations. J. Spacecr. Rockets (2011). https://doi.org/10.2514/1.50915 (Springer)

Rim, H.J.: TOPEX orbit determination using GPS tracking system. Doctoral thesis, University of Texas at Austin, Austin, USA (1992)

Robert, C.E.: Densities based upon Jacchia's 1971 models. Celest. Mech. **4**, 368–377 (1971)

Schlegel, K., et al.: Thermospheric density structures over the polar regions observed with CHAMP. Ann. Geophys. (2005)

Standish, E.M.: JPL planetary and lunar ephemerides, DE405/LE405. Jet Propulsion Laboratory, InterOffice Memorandum, IOM (1998)

Wang, X., et al.: Precise orbit determination technology based on SLR solution for HY-2 satellite (in Chinese). Eng. Sci. (2014)

Willis, P., et al.: Recent improvements in DORIS data processing at IGN in view of ITRF2008, the ignwd08 solution. In: Geodesy for Planet Earth International Association of Geodesy Symposia (2012). https://doi.org/10.1007/978-3-642-20338-1_6 (Springer)

Willis, P. et al.: DPOD2008, A DORIS-oriented terrestrial reference frame for precise orbit determination. In: Rizos, C., Willis, P. (eds.) International Association of Geodesy Symposia, p. 143. Springer, Switzerland (2013)

Zelensky, N.P., et al.: DORIS/SLR POD modeling improvements for Jason-1 and Jason-2. Adv. Space Res. (2010). https://doi.org/10.1016/j.asr.2010.05.008 (Springer)

In Situ Test of Traffic-Load-Induced Settlement of Alluvial Silt Subsoil Treated by Unslaked Lime

Qing Jin(✉), Xinzhuang Cui, Junwei Su, Tu Lu, Lei Zhang, and Zhongxiao Wang

School of Civil Engineering, Shandong University, Jinan 250061, China
{827762282,1571201036,1033871383,1392866188, 837827723}@qq.com, cuixz@sdu.edu.cn

Abstract. The soft wet alluvial silt is widely distributed material in the world. In order to improve the bearing capacity and decrease the traffic-load-induced settlement of silt subsoil, the shallow subsoil always treats with the unslaked lime. However, the mitigating effect of this ground treatment method on traffic-load-induced settlement of alluvial silt subsoil is inconclusive. Therefore, with the developed falling weight simulation equipment of traffic load, in situ tests are carried out on the natural and unslaked lime treated alluvial silt ground in the Yellow River delta of China to study traffic-load-induced settlement. Furthermore, Chai-Miur cumulative deformation model of soil is employed to numerically simulate the long-term traffic-load-induced cumulative settlement. In situ test results indicate that because unslaked lime treatment enhances wave impedance of the reinforced soil layer, the wheel-load-induced dynamic stress and excess pore water pressure in the substratum decreases. The decrease of excess pore water pressure reduces the cumulative settlement of unslaked lime treated subsoil. For short-term cumulative settlement, there are differences between tested and calculated results, but not much. The calculated results imply that after opening to traffic for 10 years, compared with the natural ground, the cumulative settlement of the unslaked lime treated subsoil reduces by about 21%, and the change of transverse slope of pavement induced by cumulative settlement decreases by 1/3. In situ test and numerical calculation results demonstrate that shallow layer treatment with unslaked lime can effectively mitigate the cumulative settlement of alluvial silt subsoil.

1 Introduction

Alluvium is loose soil, which has been eroded and reshaped by water in some form, and typically made up of a variety of materials, including fine particles of silt and clay and larger particles of sand and gravel. Alluvium is widely distributed in the world. In China, the Yellow River delta is the youngest large river delta due to frequent river diversion in history, and in the Yellow River delta, the alluvium is mainly low liquid limit silt. The alluvial silt is newly formed under consolidated deposit and has poor engineering performances with low liquid limit and plasticity index, small cohesion, low strength, intensive capillarity, poor graduation and water stability and high

© Springer International Publishing AG, part of Springer Nature 2019
A. Sevi et al. (eds.), *Enhancements in Applied Geomechanics, Mining, and Excavation Simulation and Analysis,* Sustainable Civil Infrastructures,
https://doi.org/10.1007/978-3-319-95645-9_13

liquefaction potential. Furthermore, because the alluvial silt is liquefiable and the groundwater level is very high in the region, the traffic-load-induced cumulative settlement of the road with low embankment is significant after opening to traffic, and this can cause serious pavement diseases. In order to improve the bearing capacity of the soft wet subsoil in the region and reduce its cumulative settlement, shallow layer treatment with unslaked lime is usually employed. The improving effect of unslaked lime treatment on the bearing capacity of soft wet subsoil is undeniable (Kamon 1992; Narasimha Rao and Rajasekaran 1996). However, at present the mitigating effect of hard crust formed by unslaked lime treated subsoil on the traffic-load-induced cumulative settlement is inconclusive. Note that in this study, the pavement means the hard layered structure that forms a road carriageway, the embankment means the compacted soil layer below the pavement of a road, and the subsoil means the natural soil below a road embankment; the term subgrade includes all layers above the natural ground surface.

Previous studies on the cumulative settlement of subsoil focus on dynamic deformation test of soil in the laboratory and numerical simulation. In the early 1950s, Seed and his co-workers (Seed et al. 1955; Seed and McNeill 1956) studied the settlement behaviour of road under repeated loads by normal compression tests. Monismith et al. (1975) analyzed the characteristics of permanent deformation of subsoil due to repeated loadings based on repeated-load triaxial compression tests. Yildirim and Ersan (2007) studied consolidation settlements of soft clay by undrained cyclic simple shear tests in the laboratory. Shahu and Yudhbir (2008) studied cumulative plastic strain of a quasi-saturated compacted silty clay under cyclic load by monotonic and cyclic undrained triaxial tests. Cyclic triaxial tests were conducted by Liu and Xiao (2010) to study the behaviour of silt subsoil under various physical states and stress conditions. Chai and Miura (2002) modified the Li-Selig model (Li and Selig 1996) and calculated the permanent settlement of road with a low embankment on soft subsoil. Akira et al. (2003) analyzed the traffic-load-induced settlement of low embankment road on silty-clay. Abdelkrim et al. (2003) adopted a general structure analysis approach to predict the traffic-load-induced residual settlement. Fujiwara and coworkers (Fujiwara et al. 1985; Fujiwara and Ue 1990) researched the effect of preloading on the cumulative settlement of clay under cyclic loadings, and they concluded that the amount of settlement that occurs after construction depends strongly on the soil over consolidation ratio, degree of consolidation at the time of unloading, static loading magnitude, and repeated loading magnitude. Chai and Miura (2002) calculated the permanent settlement of the improved subsoil with soil-cement columns. Indraratna et al. (2010) carried out a large-scale triaxial testing on clay-drain complex subjected to cyclic loading representing a typical track environment to predict the behaviour of the soft estuarine subsoil with short vertical drains. However, currently there has been no studies on the cumulative settlement of unslaked lime treated subsoil.

Though the laboratory tests can evaluate the traffic-load-induced cumulative deformation of subsoil, they are not accurate because the stress and boundary conditions are more complex in real practices. The numerical method, although it has the advantage of analyzing the subsoil settlement under complicated stress conditions, has its shortcomings, such as long calculating time and uncontrollable cumulative calculation errors, which limit its practical use in engineering. In addition, the road settlement observed in the field after opening to traffic is the sum of the settlement induced by the weight of subgrade and traffic-load-induced settlement. However, there is no effective method to separate them and analyze individually traffic-load-induced cumulative settlement. Therefore, it is significant to conduct in situ tests for directly simulating the traffic-load-induced settlement of subsoil. In this paper, with the developed Falling-Weight Traffic Load Simulation Equipment (FWTLE), in suit tests are performed to study the short-term traffic-load-induced cumulative settlement of unslaked lime treated alluvial silt subsoil in the Yellow River delta. And by the numerical calculations, the long-term cumulative settlement is studied. Comparing the cumulative settlement of the natural ground with that of unslaked lime treated subsoil demonstrates the validity of unslaked lime treatment method.

2 In Suit Tests

2.1 Description of the Site

The test section was constructed at the Shouguang prefecture road, the Xinhe-Xinzhuangzi line in the Yellow River delta. The topsoil is the Yellow River alluvial silt with low liquid limit, and its parameters are shown in Table 1. The groundwater level of the test site is 0.6 m below the ground surface. The alluvial silt has a low cohesion, and it is easy to be liquefied. The average saturation of the surface silt reaches 0.78 because of its strong capillarity. Its curvature coefficient of grading curve is greater than 3, so the low liquid limit silt belongs to the poorly graded soil. X-ray diffraction test showed that the non-clay mineral content is more than 80%. Microscopic structure of particles was analyzed with JXA-8800R electronic probe, as shown in Fig. 1. Figure 1a shows the common silt from Jinan pre-mountain proluvial alluvial silt with clay particles removed, and Fig. 1b shows the low liquid limit silt in the Yellow River delta. Compared with common silt, the Yellow River delta silt has high psephicity and less the elongated and flaky particles. By the long time actions of soaking and erosion in water, particles impacting and water scouring, the surface of the alluvial silt particles is broken and eroded seriously. So it is hard to be compacted.

Table 1. Geomechanical parameters of natural subgrade

Soil layer	Liquid limit (%)	Plasticity index (%)	Water content (%)	Void ratio	Saturation	Cohesion (kPa)	Internal friction angle (°)	Compression modulus (MPa)
0–0.6 m	25	7.4	17.6	0.602	0.78	31.8	22.9	7.45
0.6–10 m	27	9.8	18.	0.603	1.00	26.6	20.5	15.86

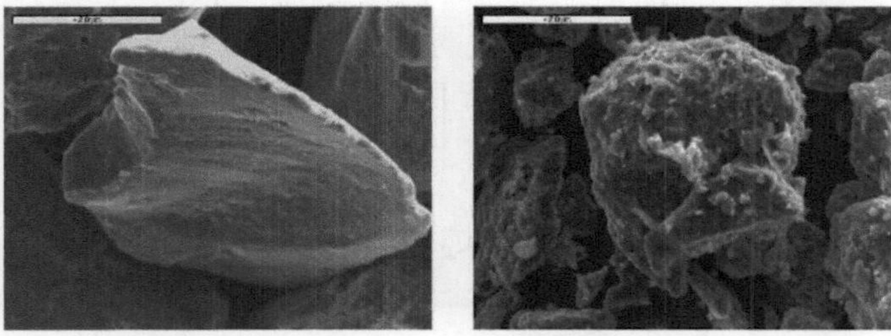

(a) Common silt **(b) TheYellow River delta silt**

Fig. 1. Microstructure of silt (2k times)

In order to improve the bearing capacity of the subsoil and decrease the traffic-load-induced cumulative settlement, the design requires that the 60 cm thick unsaturated topsoil is treated with 6% unslaked lime powder. The physical mechanical properties of unslaked lime treated soil are shown in Table 2.

Table 2. Physical and mechanical parameters of unslaked lime treated soil

Lime content	Optimum moisture content (%)	Maximum dry density (g/cm^3)	Elastic modulus (MPa)	Cohesion (kPa)	Internal friction Angle (°)
6%	15	1.81	77	115.4	34.7

2.2 Test Devices and Parameter Settings

In order to simulate the action of wheel load on the subsoil, the FWTLE was developed to simulate the cumulative settlement of natural ground at the test site (Cui 2012). The FWTLE is composed of three parts: automatic control system, pneumatic systems and loading system, as shown in Fig. 2.

For in situ simulated test of cumulative settlement, the parameter settings of the FWTLE are critical. These parameters include the weight and height of falling weight, the stiffness of air spring and the size of loading plate. By setting different parameters, the different stress response of subsoil can be simulated.

The height of the road embankment has a great effect on traffic-load-induced vertical stress on the ground surface. According to the earlier investigation (Cui 2012), for the vehicle with standard axle load of 100 kN, the common highway pavement structure (the thickness is approximately 70 cm) and zero-fill embankment (the height of the embankment is 0 m), the vertical stress on the ground surface is about 22 kPa. But when the height of the embankment is 0.8 and 1.5 m, the vertical stresses reduce to 7.8 and 5.6 kPa, respectively. Simultaneously, for zero-fill embankment, when the horizontal distance from the center of wheel gap is more than 0.7 m, the vertical stress changes slowly with the horizontal distance, and this is basically irrelevant to the action

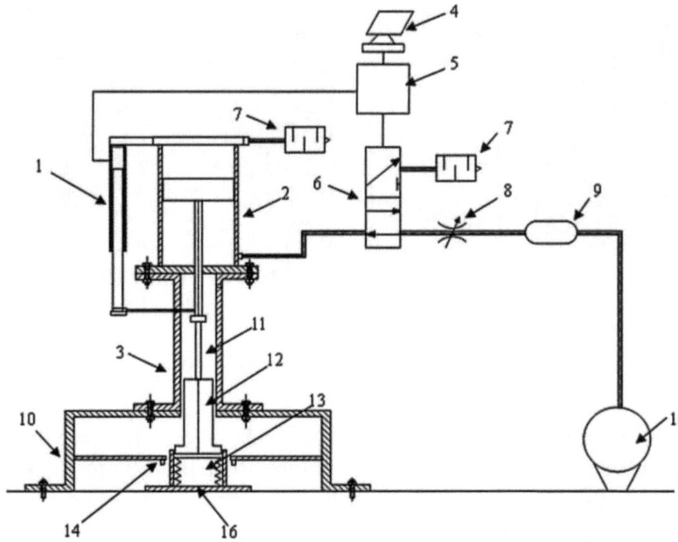

1-Displacement sensor; 2-Cylinder; 3-Guiding device; 4-Computer; 5-Single chip; 6-Two-position three-way electromagnetic value; 7-Silencer; 8-Throttle valve; 9-Stable pressure box; 10-Supports; 11-Lifting rope; 12-Drop hammer; 13-Air spring; 14-Laser displacement meter; 15-Air pump; 16-Loading plate

Fig. 2. Schematic overview of in situ simulation equipment of traffic load

time of wheel load (i.e. vehicle speed). Therefore, the size of square loading plate used in situ test is determined as 1.2 m × 1.2 m, and its thickness is 3 mm. And the radius of its equivalent area circle is 0.68 m, which can meet the test precision requirement in the case of the zero-fill embankment. The falling weight in the test is the hammer used in the standard penetration test (SPT) for the engineering geological investigation and it is 62.5 kg in weight. Before in situ test, by adjusting the height of falling weight and the stiffness of air spring, different vertical stress under the loading plate can be obtained to meet the requirements of different engineering cases. The amplitude and duration of vertical dynamic stress response obtained in tests should be approximately consistent with the computed results in the cases of different heights of the embankment and different vehicle speeds.

Because the cumulative settlement of the zero-fill road embankment is significantly larger, in this study, the falling distance and the stiffness of air spring were adjusted to simulate the case of the zero-fill embankment.

2.3 Layout of Sensors

Before in situ tests, remove weeds and cover soil on the ground. In order to study the mitigating effect of unslaked lime treatment method on the cumulative settlement, natural and unslaked lime treated subsoil were tested, respectively. In the process of cyclic loading, the parameters such as stress, displacement and pore pressure were tested. Because the vehicle-load-induced excess pore pressure need a long time to dissipate, the settlement and pore pressure after cyclic loading were also tested. The total cumulative settlement was the sum of the settlement during and after loading.

The subsoil settlement was tested with the laser displacement sensors. Four dynamic soil pressure sensors were installed at different characteristic positions between loading plate and subsoil surface. Two dynamic pore pressure sensors were installed at 0.7 and 1.2 m underground, respectively, below the center of loading plate. And in the same way, two dynamic soil pressure sensors were installed at 0.6 and 1.0 m underground, respectively (Fig. 3).

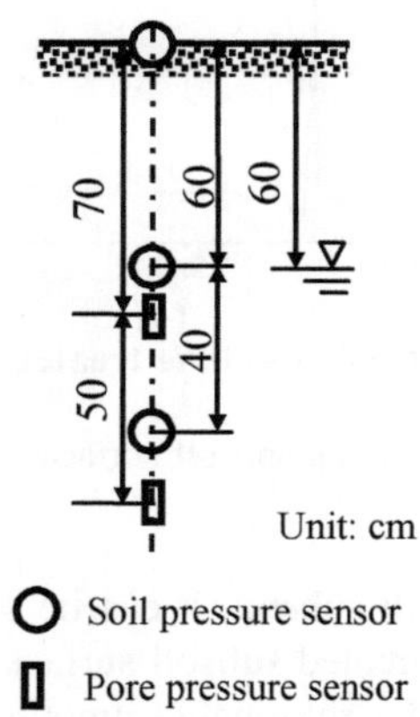

Fig. 3. Layout diagram of sensors

2.4 Test Results and Analyses

2.4.1 Dynamic Stress

Figure 4 shows vertical stress response curves tested by dynamic soil pressure sensors on the subsoil surface under the center of the loading plate. In tests, the falling distance of weight is 10 cm. It can be seen from Fig. 4a, the time-histories of stress response basically agree with the factual vertical dynamic stress response of the natural ground surface under a vehicle load (Wang 2007). For natural subsoil, the tested vertical stress amplitude on the ground is 23 kPa. As mentioned above, for the common pavement structure model, under the action of standard axle load of 100 kN (double-wheel and single-axle load stated in the Chinese specifications for design of pavement), the vertical stress amplitude on the ground surface under the zero-fill embankment is 22 kPa. The tested and calculated stress amplitudes have little difference. In addition, the single loading period of the dynamic stress corresponding to the 10 cm height of falling weight shown in Fig. 3a is 0.031 s, which is approximately equivalent to the vehicle speed of 120 km/h (Huang 1993). Therefore, when the falling distance is 10 cm, the FWTLE can simulate the stress induced by the moving vehicle with the speed of 120 km/h on the road with zero-filled embankment.

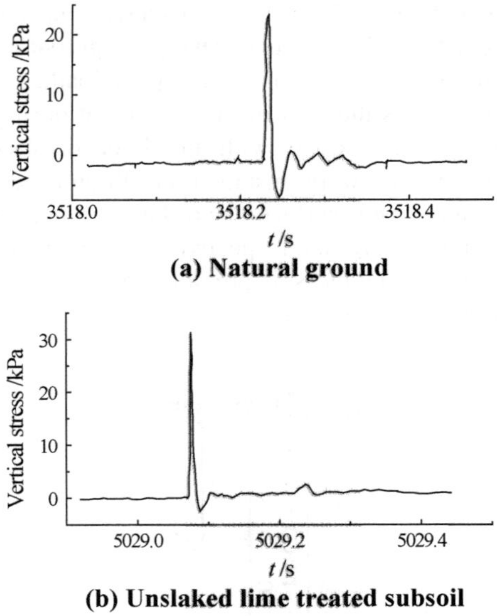

Fig. 4. Vertical stress response on the subsoil surface under the center of loading plate

Compared with the natural silt subsoil, it can be seen from Fig. 4b that the stress amplitude on the unslaked lime treated subsoil surface increases. This is because after mixing the soft wet soil with the reasonable amount of unslaked lime, a series of reactions take place: water absorption, exothermic action and expansive action; ion exchange action; carbonation (chemical cementation reaction); pozzolanic action (chemical gelation reaction) and crystallization action. These reactions can make the moisture in soil reduce, slit particles coagulate to form larger aggregates. With the gradual hardening of subsoil, soil particles bonded together and the physical and mechanical properties of the subsoil are improved. With the subsoil treated by unslaked lime, the wave impedance of shallow subsoil increase, and the hard curst effect causes the increase of the dynamic stress amplitude on the subsoil surface.

Figure 5 shows that the variation curves of vertical dynamic stress amplitude of the subsoil under the center of the loading plate with depth. It is can be seen that compared with the natural ground, the vertical stress amplitude in the ground treated by unslaked lime more quickly decreases, and this makes the dynamic stress in the substratum of hard crust significantly reduce. The decrease of the stress level in the substratum can induce the reduction of the cumulative deformation.

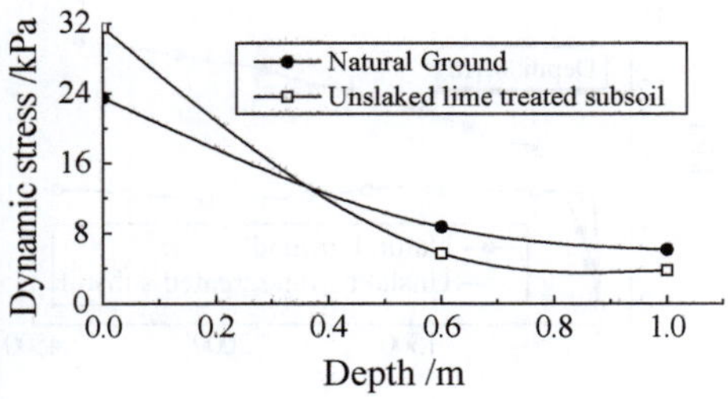

Fig. 5. Variation curves of vertical stress amplitude in subsoil with depth

2.4.2 Excess Pore Water Pressure

The repeated traffic load induces the excess pore water pressure in the subsoil, and excessive pore water pressure can cause the liquefaction of silt subsoil. Tests reveal that, as the number of cyclic load increases, cracking and mud pumping can be seen on the ground surface (Fig. 6), and this indicates the subsoil has been liquefied.

Fig. 6. Mud-pumping of subsoil

Figure 7 shows the variations of the excess pore water pressures in the natural and unslaked lime treated ground with the load numbers N during the loading process. It can be seen that, compared with the natural ground, the excess pore water pressure in the soft substratum of the unslaked lime treated ground is obviously lower. This is because with the ground treated by unslaked lime, the dynamic stress level in the substratum decreases (as shown in Fig. 5).

At the initial stage of loading, the pore water pressure in the soft substratum of the unslaked lime treated ground develops faster than that in the natural ground. Nevertheless, after certain load numbers, the pore water pressure in unslaked lime treated ground does not increase but gradually declines to a stable state at last. By contrast, the pore water pressure in natural ground continuously increases with the load numbers.

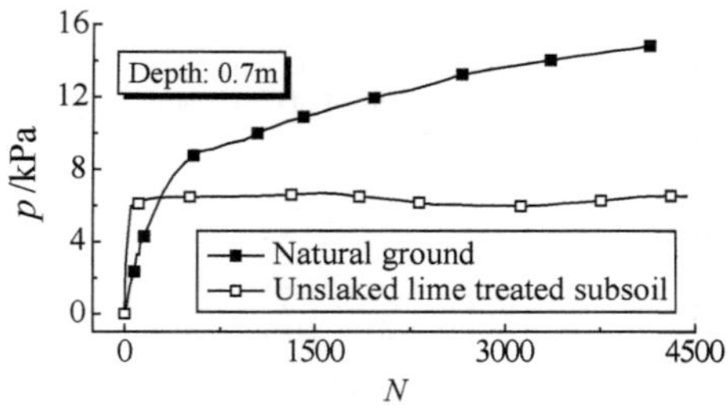

(a) 0.7 m depth under ground

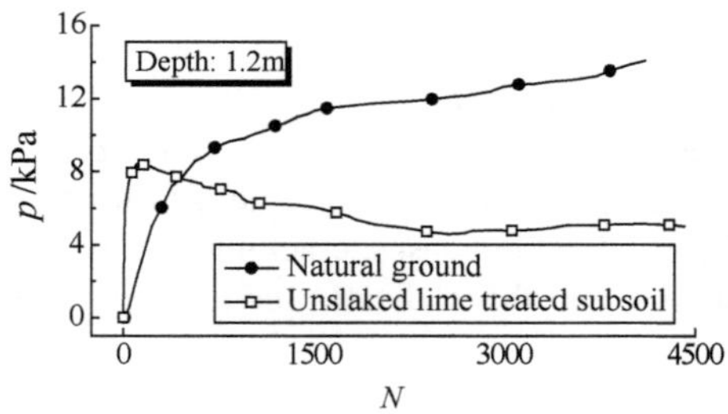

(b) 1.2 m depth under ground

Fig. 7. Variation curves of excess pore water pressure with load numbers

The difference above is mainly because of the different drainage conditions of the unslaked lime soil treated and natural grounds during the loading process. Because the permeability coefficient of unslaked lime treated soil is smaller than the natural soil, the soil-lime hard crust can block the dissipation of pore water pressure in the substratum. Therefore, at the beginning of cyclic loading, the pore water pressure in unslaked lime treated ground increases faster than the one in the natural ground. But as the pore water pressure increases, the strength of the substratum decreases, and this induces the cracking of the soil-lime hard crust under the traffic load (observed in tests). The cracks provide good drainage channels. As mud comes out through the channels, the pore water pressure in the substratum of hard crust does not increase, but decreases. However, for the natural ground, the flexibility of top soil is larger, so its main failure mode is not local cracking but the plastic deformation. Therefore, the seepage paths change little in the process of cyclic loading, and this makes the pore water pressure continuously increase.

Figure 8 shows the variations of excess pore water pressure in natural and unslaked lime treated grounds after terminating loading. The excess pore water pressure sharply declines firstly, and then stabilizes for some time, finally declines to hydrostatic pressure. This phased variation implies that after terminating loading, the seepage path of water and dissipation process of pore water pressure are complex.

After terminating loading, compared with the natural ground, the dissipation of excess pore water pressure in unslaked lime treated ground needs more time. The dissipation of excess pore water pressure in natural ground needs seven hours, but the one in unslaked lime treated ground takes more than twenty hours. This is because after

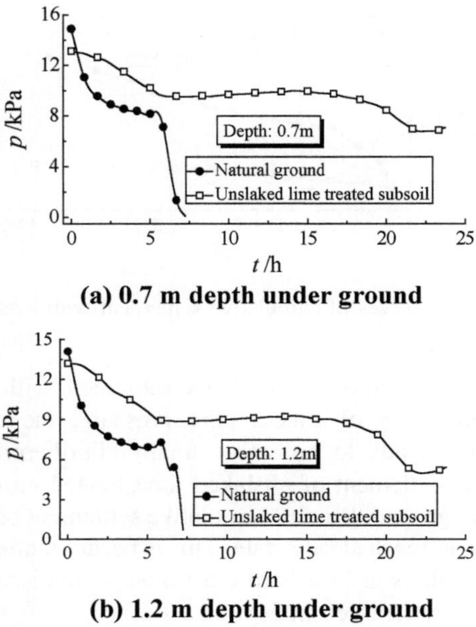

Fig. 8. Dissipation of excess pore water pressure after terminating loading

terminating cyclic loading, the mud hardens in the cracks of soil-lime hard crust induced by cyclic loading, and this makes the hard crust becomes impervious. However, the practical traffic load is cyclic and continual, the cracks in hard crust are not blocked. So the delay effect of dissipating the pore water pressure does not occur.

2.4.3 Cumulative Settlement

The traffic-load-induced cumulative deformation of soil is mainly composed by undrained shear deformation and consolidation deformation. Under practical traffic loads, these two deformations occur simultaneously. However, in model tests, only in the process of cyclic loading, are these two deformations concurrent. After terminating loading, the settlement of subsoil is mainly induced by consolidation deformation of soil. Although the deformation paths of subsoil under the practical and simulated vehicle load are different, the sum of the undrained shear and consolidation deformations obtained in the tests can be regarded as the total cumulative deformation induced by traffic load (Cui 2012).

Figure 9 shows the variation curves of the cumulative settlement with the load numbers. It can be seen that the growth of cumulative settlement gets slow gradually with the increase of load numbers. After the subsoil is treated with unslaked lime soil, the cumulative settlement obviously decreases. This illustrates that, for the Yellow River alluvial silt, the unslaked lime treatment can effectively mitigate the traffic-load-induced cumulative settlement.

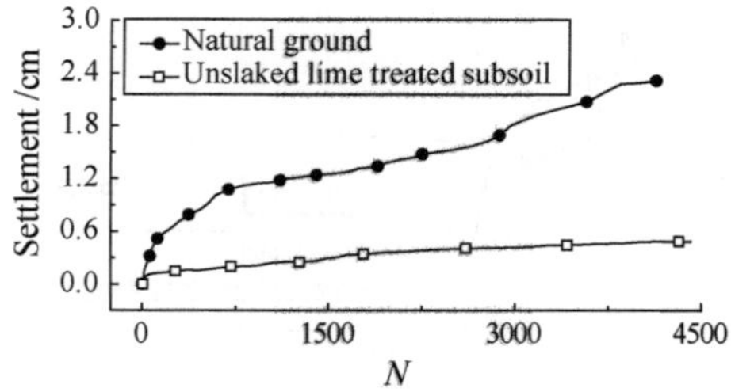

Fig. 9. Variation curves of cumulative settlement with loading numbers

Figure 10 shows the variations of cumulative settlement with time after terminating loading. With the dissipation of excess pore pressure, the cumulative settlement gradually increases, and tends to be stable finally. Furthermore, after terminating loading, the cumulative settlement of unslaked lime treated ground has a little difference with that of natural ground. But the cumulative settlement of unslaked lime treated subsoil takes more time to reach steady state. This is because, after terminating loading, the hardening of mud in the soil-lime hard crust blocks the dissipation channels of the excess pore water pressure (as seen in Fig. 8).

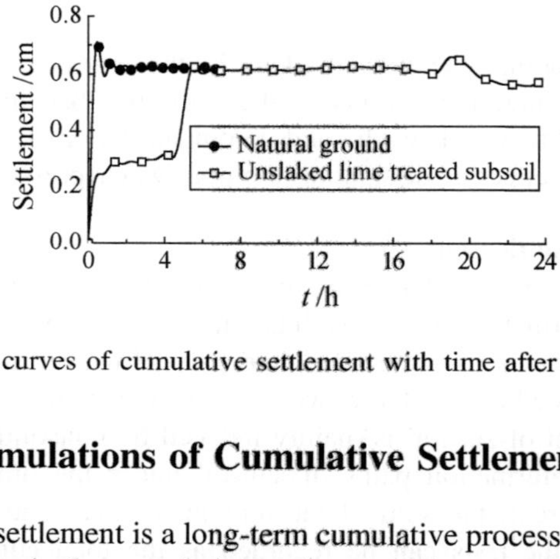

Fig. 10. Variation curves of cumulative settlement with time after terminating loading

3 Numerical Simulations of Cumulative Settlement

Traffic-load-induced settlement is a long-term cumulative process. In situ tests as above can be employed to study the cumulative settlement. However, its conduction is costly and time-consuming, so it is only suited for studies of the initial fast-developing cumulative settlement. Therefore, numerical method is widely employed to predict the traffic-load-induced long-term cumulative settlement. At the early age, Monismith et al. (1975) proposed a power model of cumulative deformation of soil, and then it was modified by Li and Selig (1996). Chai and Miura (2002) made further improvement to Li-Selig model. The Chai-Miura model considered not only the dynamic deviatoric stress and static strength of soil but also the effect of initial static deviatoric stress on cumulative deformation. The Chai-Miura model is widely used to calculate the cumulative deformation of soil.

Chai-Miura model is expressed as follows:

$$\varepsilon_{\mathrm{p}} = a\left(\frac{q_{\mathrm{d}}}{q_{\mathrm{f}}}\right)^{m}\left(1+\frac{q_{\mathrm{s}}}{q_{\mathrm{f}}}\right)^{n} N^{b} \tag{1}$$

where $q_{\mathrm{s}} = \sqrt{3J_{2\mathrm{s}}}$ is the initial static deviatoric stress, $J_{2\mathrm{s}}$ is the second deviatoric stress invariant of initial static stress; $q_{\mathrm{d}} = \sqrt{3J_{2\mathrm{d}}}$ is the dynamic deviatoric stress, $J_{2\mathrm{d}}$ is the second deviatoric stress invariant of dynamic stress peaks in all directions; q_{f} is the static strength of soil; *a, b, m* and *n* are parameters of soil; *N* is the load numbers.

According to the effective consolidation stress theory (Shen 2000), the static strength of soil q_{f} can be determined by strength index c_{cu} and φ_{cu} of the consolidation undrained total stress:

$$q_{\mathrm{f}} = c_{\mathrm{cu}} \cos\phi_{\mathrm{cu}}/(1-\sin\phi_{\mathrm{cu}}) + \frac{1}{2}(1+K_0)\sigma_{\mathrm{cz}}\sin\phi_{\mathrm{cu}}/(1-\sin\phi_{\mathrm{cu}}) \tag{2}$$

where K_0 is the static soil pressure coefficient at rest; σ_{cz} is overlying soil pressure.

a, b, m and *n* in Eq. (1) reflect the combined effect of stress state, physical state and types of soil. Series of triaxial tests, the constants in Eq. (1) were obtained: $a = 0.64$, $b = 0.10$, $m = 1.70$ and $n = 1.00$.

3.1 Comparisons and Analyses of the In Situ Test and Numerical Results

Before calculating long-term cumulative settlement of subsoil, in order to prove the validity and feasibility of calculation method, the cumulative settlement of natural ground was numerically calculated by simulating in situ test.

Firstly, the static module of the finite difference program Flac3D was used to calculate initial static deviatoric stress q_{s} in the subsoil. Secondly, the dynamic module of Flac3D was employed to simulate the dynamic response of subsoil. In the dynamic calculation, the vertical dynamic stresses obtained by soil pressure sensors (Fig. 3a) under the loading plate in the tests were loaded on the subsoil surface. The dynamic deviatoric stress q_{d} can be obtained from the peaks of the dynamic stresses in three orthogonal directions. Finally, q_{s} and q_{d} were taken into Eq. (1), and cumulative deformation of subsoil can be calculated. The static strength of soil q_{f} is calculated by Eq. (2). The cumulative settlement on the subsoil surface can be obtained by integrating the vertical cumulative deformation of soil along depth.

Figure 11 shows the comparisons of cumulative settlements from in situ tests and numerical simulations. It can be seen that the variation trend of testing cumulative settlement with the load numbers is basically consistent with calculated results. However, the calculated settlement develops faster than the tested results at the early stage of cyclic loading. There are many reasons for the differences between the tested and calculated settlements. The main reason is that the subsoil is partially drained at the site, i.e. the development and dissipation of excess pore water pressure in the subsoil

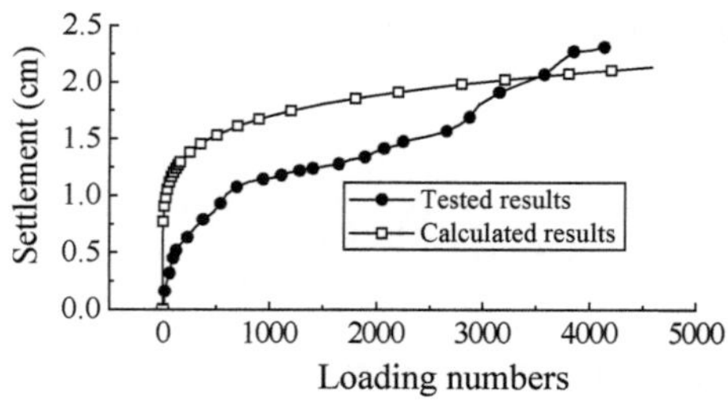

Fig. 11. Comparisons between calculated and tested cumulative settlements

are simultaneous in the process of loading, however, Chai-Miura model used in numerical simulation was established based on undrained shear tests.

3.2 Numerical Simulations and Analyses of Long-Term Cumulative Settlement

Long-term cumulative settlement of the subsoil of Xinhe-Xinzhuangzi expressway is studied. This expressway is located in the Yellow River delta with bidirectional four lanes and the average embankment height of 1.4 m. The road structure and material parameters are shown in Table 3.

Table 3. Road structure and material parameters

Structural layer	Materials	Thickness (cm)	Elastic modulus (MPa)	Poisson ratio
Upper surface	SMA	4	1400	0.3
Middle surface	Mesograin modified asphalt concrete	6	1200	0.3
Lower surface	Coarse graded asphalt concrete	8	1000	0.3
Upper base	Large grain size asphalt gravel	12	1400	0.35
Lower base	Cement stabilized gravel	36	1500	0.35
Subbase	Lime-ash soil (30% additive gravel)	20	800	0.35
Roadbed	Low liquid limit silt (96% compaction degree)	80	30	0.35
Embankment	Low liquid limit silt (94% compaction degree)	60	17	0.35

Referring to the Chinese specification of General Code for Design of Highway Bridges and Culverts, the load class of truck-20 (the live load of 200 kN) was employed to load in the middle of the carriageway. The weights of the front and rear axles were 70 and 130 kN, respectively. In the calculation, the wheel loads were

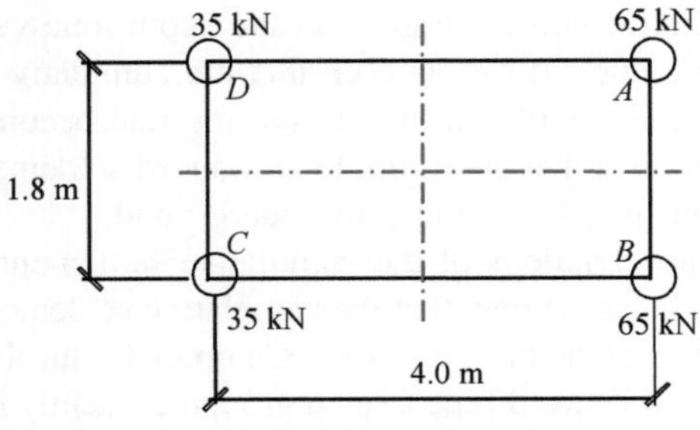

Fig. 12. Sketch map of truck-20 load

simplified as four point loads on the pavement shown in Fig. 12. The loading mode of wheel load proposed by Huang (1993) is adopted:

$$\begin{cases} F = F_{\max} \sin^2\left(\frac{\pi}{T_s} t\right) & 0 \leq t \leq T_s \\ F = 0 & t > T_s \end{cases} \tag{3}$$

where t is time; $F_{\max}$ is the wheel load peak, 65 kN for the front wheel and 35 kN for the rear wheel; T_s is the duration time of single vehicle load, and has an inverse relationship with the vehicle speed. Herein T_s is taken as 0.031 s, representing the equivalent vehicle speed of 120 km/h (Huang 1993).

The settlement calculation considers the variation of annual traffic volume. The traffic volumes at eight flyovers along the Xinhe-Xinzhuangzi highway were investigated. The predicted cumulative traffic volume of trucks is:

$$N = \frac{73N_1}{\gamma}\left[(1+\gamma)^t - 1\right] \tag{3}$$

where N_1 = 13463 is the annual average daily traffic at the early stage after opening to traffic; r = 5.775% is the average annual growth rate; t is time in year.

The physical and mechanical parameters of pavement, embankment and subsoil are shown in Tables 1, 2 and 3. In this study, for the unslaked lime treated ground, the cumulative settlement is only from the deformation of the natural soil layer, and the deformation of soil-lime hard crust was ignored in the calculation. Figure 13 shows

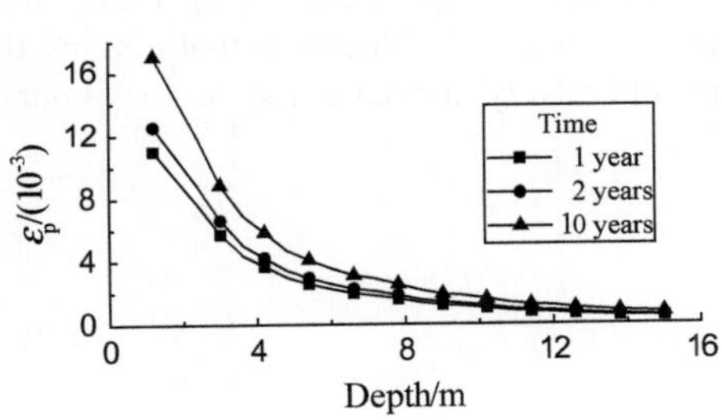

Fig. 13. Variation curves of cumulative deformation with depth

the variations of cumulative deformations with the depth (one year, two years and ten years after opening to traffic). It can be seen that the cumulative deformation sharply reduces within 5 m underground, and the decreasing rate becomes slow beyond 5 m underground. This illustrates that the traffic-load-induced settlement is mainly from the cumulative deformations of silts within 5 m underground.

Figure 14 shows the variations of the cumulative settlement of subsoil with time after opening to traffic. It can be seen that the cumulative settlement rapidly develops in the initial stage and then gradually gets slow. Compared with the natural ground, the cumulative settlement of unslaked lime treated ground evidently reduces. For example, the cumulative settlement in ten years reduces by 21.4%. This demonstrates that the unslaked lime treatment obviously mitigates the traffic-load-induced settlement, and this is consistent with the results of in situ tests.

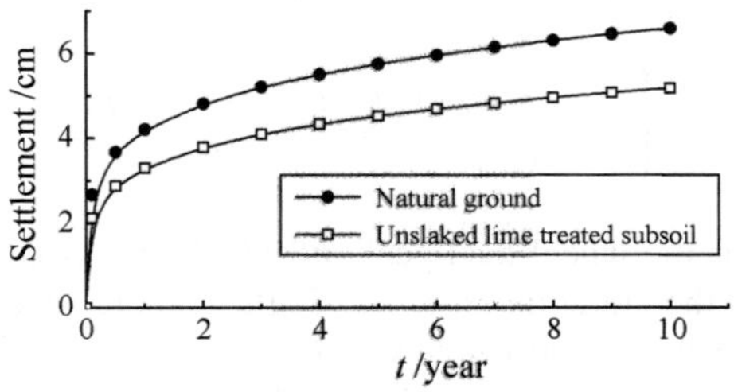

Fig. 14. Variation curves of cumulative settlement with time

Figure 15 depicts the transverse distribution of cumulative settlement after opening to traffic for 10 years. It can be seen that the cumulative settlement of the carriageway is the largest, and this can induce an additional transverse slope on the pavement which is opposite to the designed road camber for pavement transverse drainage. According to the Chinese specification of code for design of urban road engineering, for high class pavement, the average transverse slope of the road crown is 1–2%. However, the traffic-load-induced cumulative settlement greatly leads to the decrease of the transverse slope. For natural ground, the cumulative settlement causes the inward transverse slope on the outside of the road is 0.47%. This can lower drainage performance of road and induce surface water, which not only affects the driving safety but also exacerbates the destruction of the pavement. However, for the unslaked lime treated ground, the change of the outside transverse slope induced by traffic load is reduced by 1/3 compared with the natural ground. This illustrates that treating the soft wet alluvial silt with unslaked lime can effectively decrease the harms induced by the cumulative settlement.

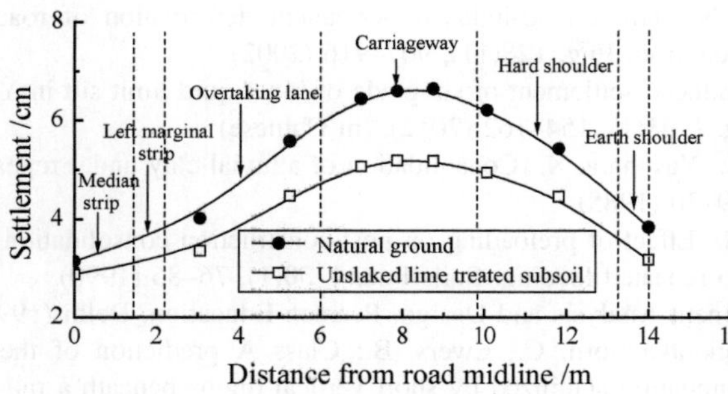

Fig. 15. Transverse distribution curves of cumulative settlement after opening to traffic for 10 years

4 Conclusions

In order to prove the mitigating effect of unslaked lime treatment on the traffic-load-induced settlement of soft wet alluvial silt subsoil, in situ tests were conducted to study the short-term settlements of natural and treated grounds in the Yellow River delta with the developed FWTLE. In addition, based on the Chai-Miura cumulative deformation model, the long-term cumulative settlements of the subsoils were analyzed numerically. The following main conclusions are drawn out:

(1) Compared with the natural ground, the wave impendence of unslaked lime treated ground increases. This makes traffic-load-induced dynamic stress in substratum significantly reduce. Moreover, the excess pore water pressure in the substratum of unslaked lime treated subsoil is less than that in the natural ground.
(2) In comparison with the natural ground, the cumulative settlement of the unslaked lime treated ground obviously decreases. This proves the mitigating effect of unslaked lime treatment on the cumulative settlement of alluvial silt subsoil.
(3) The cumulative settlement of subsoil aggravates the development of pavement diseases. Unslaked lime treatment of subsoil can effectively decrease the harms of traffic loads on the road with low embankment.

Acknowledgements. This work was supported by the Chinese Natural Science Foundations (Nos. 51279094, 51078222 and 50708056), the Natural Science Foundations of Shandong Province, China (No. ZR2011EEM012) and the Independent Innovation Foundation of Shandong University (IIFSDU) (No. 2012HW003).

References

Abdelkrim, M., Bonnet, G., Buhan, P.D.: A computational procedure for predicting the long term residual settlement of a platform induced by repeated traffic loading. Comput. Geotech. **30**(6), 463–476 (2003)

Akira, S., Lawalenna, S., Norihiko, M.: Partially-drained cyclic behavior and its application to the settlement of a low embankment road on silty-clay. Jpn. Geotech. Soc. **43**(1), 33–46 (2003)

Chai, J.C., Miura, N.: Traffic-load-induced permanent deformation of road on soft subsoil. J. Geotech. Geoenviron. Eng. **128**(11), 907–916 (2002)

Cui, X.Z.: Traffic-induced settlement of subgrade of low liquid limit silt in Yellow River delta. China Civil Eng. J. **45**(1), 154–162 (2012). (in Chinese)

Fujiwara, H., Ue, S., Yasuhara, K.: Consolidation of alluvial clay under repeated loading. Soils Found. **25**(3), 19–30 (1985)

Fujiwara, H., Ue, S.: Effect of preloading on post-construction consolidation settlement of soft clay subjected to repeated loading. Soils Found. **30**(1), 76–86 (1990)

Huang, Y.H.: Pavement Analysis and Design. Pearson Education, Delhi (1993)

Indraratna, B., Rujikiatkamjorn, C., Ewers, B.: Class A prediction of the behavior of soft estuarine soil foundation stabilized by short vertical drains beneath a rail track. J. Geotech. Geoenviron. Eng. **136**(5), 686–696 (2010)

Kamon, M.: Recent developments of soil improvement. In: Proceedings of International Symposium on Soil Improvement and Pile Foundation, Nanjing, China, vol. I, pp. 1–16 (1992)

Li, D., Selig, E.T.: Cumulative plastic deformation for fine-grained subgrade soils. J. Geotech. Eng. **122**(12), 1006–1013 (1996)

Liu, J.K., Xiao, J.H.: Experimental study on the stability of railroad silt subgrade with increasing train speed. J. Geotech. Geoenviron. Eng. **136**(6), 833–841 (2010)

Monismith, C.L., Ogawa, N., Freeme, C.R.: Cumulative deformation characteristics of subsoil due to repeated loading. Transp. Res. Rec. **537**, 1–17 (1975)

Narasimha, R.S., Rajasekaran, G.: Reaction products formed in lime-stabilized marine clays. J. Geotech. Eng. **122**, 329–336 (1996)

Seed, H.B., Chan, C.K., Monismith, C.L.: Effects of repeated loading on the strength and deformation of compacted clay. Highw. Res. Board Proc. **34**, 541–558 (1955)

Seed, H.B., McNeill, R.L.: Soil deformation in normal compression and repeated loading test. Highw. Res. Board Bull. **141**, 44–53 (1956)

Shahu, J.T., Yudhbir, Hayashi S.: Cumulative plastic strain and threshold stress of a Quasi-saturated compacted silty clay. Lowland Technol. Int. **10**(2), 10–20 (2008)

Shen, Z.J.: Earth pressure of clay based on effective consolidation stress theory. Chinese J. Geotech. Eng. **22**(3), 353–356 (2000). (in Chinese)

Wang, X.: Test on dynamic stress of roadbed and pavement under heavy loads. J. Vib. Shock **26** (2), 169–173 (2007). (in Chinese)

Yildirim, H., Erşan, H.: Settlements under consecutive series of cyclic loading. Soil Dyn. Earthq. Eng. **27**, 577–585 (2007)

Experimental Study on Shear Strength Behavior of Glass Fiber-Reinforced Sand

Suchit Kumar Patel(✉) and Baleshwar Singh

Department of Civil Engineering, Indian Institute of Technology Guwahati, Guwahati, Assam, India
{p.suchit,baleshwar}@iitg.ac.in

Abstract. An experimental program was undertaken to study the effects of glass fiber reinforcement on the shear strength characteristics of a locally available sand. Fibers up to 4% was mixed uniformly with the sand and then compacted into triaxial test specimens having soil relative densities ranging from 35 to 85%, prior to testing. The effects of fiber content, soil relative density and confinement pressure were investigated. The results indicate that with the increase of relative density, the stiffness, peak stress and residual stress of the reinforced sand have also increased. The fibers have induced significant apparent cohesion to the sandy soil up to 3% fiber content. This has led to increase in energy absorption capacity indicating improvement of bearing capacity of the soil. The failure modes of the specimens indicate that the fibers have restricted the bulging and shearing zones of specimens compacted at 35 and 85% relative density, respectively.

1 Introduction

Any application of soil in geotechnical construction requires its strength characterization. Natural soils when used in earthworks can be strengthened by adding continuous reinforcement inclusions (sheets, strips, or bars) in desired directions in a defined pattern, or by mixing discrete fibers randomly within the soil mass. Though randomly distributed fiber-reinforced soils can maintain strength isotropy, this method is less widely used in the field.

Benson and Khire (1994) reported that high-density polyethylene strips with different lengths and proportions can improve the strength, secant modulus and deformation behavior of sandy soils substantially. Tingle et al. (1999) found that soil-fiber composites effectively improve the strength of sandy roadway soils and the improvement is found to be dependent on fiber fraction by weight up to a limit. The strength and deformation behavior of fiber-reinforced sandy soil is governed by soil characteristics (gradation, shape, and size), fiber properties (content, aspect ratio, and modulus) (Michalowski and Cermak 2003; Lirer et al. 2011), and interfacial mechanical interaction between reinforcement and soil matrix (Tang et al. 2010). Fiber-reinforcement reduces brittleness of sandy soil (Yetimoglu and Salbas 2002), and partial suppression of dilation due to fiber-reinforcement increases effective confinement pressure which improves soil strength and settlement behavior of sandy soil (Consoli et al. 2009). Improvement in shear strength is more pronounced at low normal

© Springer International Publishing AG, part of Springer Nature 2019
A. Sevi et al. (eds.), *Enhancements in Applied Geomechanics, Mining, and Excavation Simulation and Analysis*, Sustainable Civil Infrastructures,
https://doi.org/10.1007/978-3-319-95645-9_14

stress for fiber-reinforced soil (Falorca and Pinto 2011). The effect of initial soil density is not significant on the shear strength of fiber-reinforced soil reinforced with comparatively high fiber content (Li and Zornberg 2013). Addition of polypropylene fibers to cemented soil increases peak strength, residual shear strength and energy absorption capacity but it reduces the initial stiffness and brittleness index (Hamidi and Hooresfand 2013). Tensile strength of fiber-reinforced soil increases with an increase in dry density and decreases with an increase in water content (Li et al. 2014). Fiber-reinforcement reduces the amount of particle breakage of sand, more significantly for well graded sand (Pino and Baudet 2015). Dry fiber reinforced sand provides same shear strength that of heavily compacted unreinforced moist sand (Eldesouky et al. 2016). Claria and Vettorelo (2016) found that the addition of fibers increases the shear strength and ductility but decreases the initial stiffness of sand. The shear strength was increased indefinitely with fiber length.

The objective of the present study is to investigate the influence of glass fibers on the shear strength characteristics of a locally available sand at varying relative density for use as a construction material.

2 Materials and Methods

The experimental study was conducted with sand collected from the deposits of the nearby Brahmaputra River. The sand can be designated as SP as per the Unified Soil Classification Systems (USCS). The specific gravity of sand was 2.69 with a uniformity coefficient of 1.59 and coefficient of curvature of 0.97. The maximum and minimum void ratios of the sand were 0.96 and 0.63, respectively. Glass fiber of 0.15 mm diameter, 10 mm length, 1.5 GN/m^2 tensile strength, and 2.57 specific gravity was used as reinforcement. Varying fiber contents (0.5, 1, 2, 3, and 4% by dry weight of soil) were mixed with the sand along with 2% water content addition for producing a homogeneous mix.

Prior to triaxial tests, compacted sand-fiber specimens of 38 mm diameter and 76 mm height were prepared on the base plate of the triaxial cell by using a split mould at three different relative densities (D_r = 35, 65 and 85%) of the sand. Unreinforced sand specimens were also prepared in a similar manner. Consolidated drained triaxial tests were conducted on all specimens as per ASTM D4767-11 (2011) under different confinement pressures (σ_3) ranging from 100 to 400 kPa which reflect the ranges typical of relevant geoengineering applications. Shearing was done with a strain rate of 0.12 mm/min.

3 Results and Discussions

3.1 Stress-Strain Behavior

Figures 1, 2 and 3 present the effect of fiber content on the stress-strain behavior of the sand at different relative densities under 100 kPa confinement pressure. The stress-strain plots show clear improvement in shear strength with fiber reinforcement at all

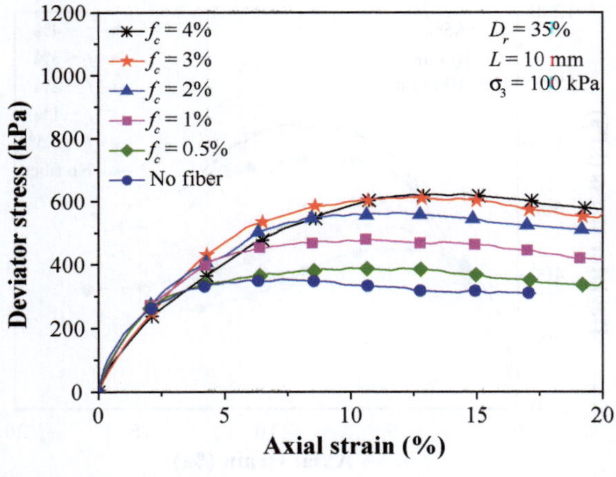

Fig. 1. Effect of fiber content on stress-strain behavior of sand (D_r = 35%)

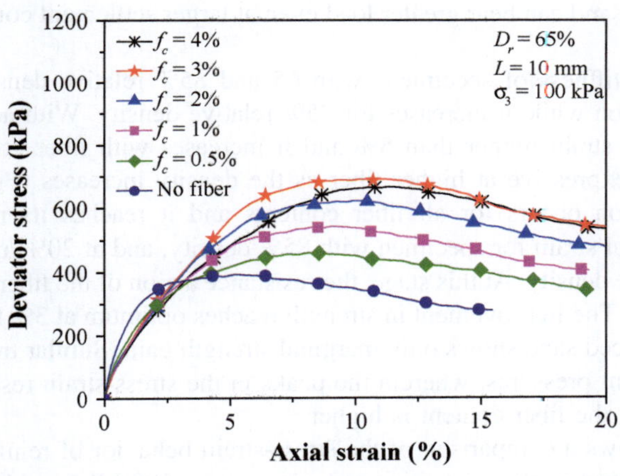

Fig. 2. Effect of fiber content on stress-strain behavior of sand (D_r = 65%)

relative densities. There is also an increase in failure strain indicating that the ductility of the reinforced sand has improved. The reinforcing action of the fibers is primarily due to local stitching of the potential slip planes through interlocking resistance.

The stress-strain plots of specimen with 35% relative density (Fig. 1) have not shown any clear peak at any fiber content. However for reinforced specimens compacted at 85% relative density (Fig. 3), the stress-strain plots show clear peak even at low fiber content (0.5%). The reinforced specimen with 65% relative density has shown an intermediate behavior (Fig. 2). In addition, the plots of the reinforced sand in the post-yield stage have exhibited strain-softening behavior and this behavior is more prominent with increasing specimen density. However, the residual strength is still greater than that of the sand alone at any density level. This indicates that the glass

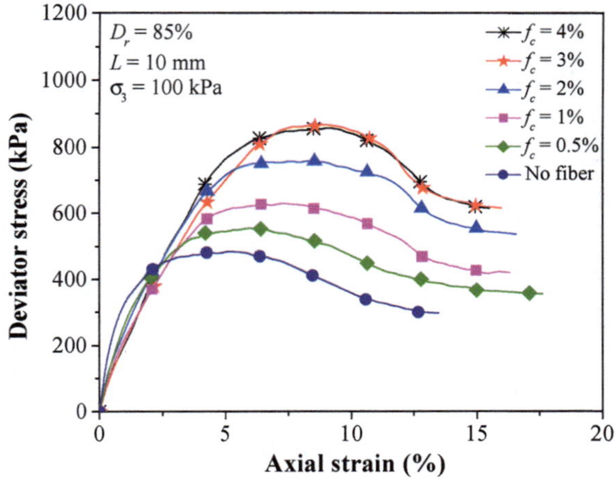

Fig. 3. Effect of fiber content on stress-strain behavior of sand (D_r = 85%)

fiber-reinforced sand can bear greater load even at larger settlement compared to parent soil.

The initial stiffness of specimens with 65 and 85% relative density is decreased with fiber addition while it increases for 35% relative density. With addition of fibers, peak appears at strain higher than 5% and it increases with fiber content. The peak becomes more expressive at higher fiber as the density increases (Fig. 3). Post-peak strength reduction occurs for all fiber contents and it reaches its residual strength around 15% axial strain for specimen with 85% density, and at 20% axial strain for 35 and 65% relative density. At this stage, the resistance action of the fibers is lost partially through pullout. The improvement in strength reaches optimum at 3% fiber content and 4% fiber-reinforced sand shows only marginal strength gain. Similar trends were found at all confinement pressures, wherein the peaks in the stress-strain responses are more prominent when the fiber content is higher.

Figure 4 shows a comparison of the stress-strain behavior of reinforced sand with 3% fiber at different relative densities under low (σ_3 = 100 kPa) and high confinement pressure (σ_3 = 400 kPa). The initial stiffness of the reinforced sand increases with either the increase of soil relative density or confinement pressure keeping the other constant indicating that the specimen density has major role in stiffness improvement. However, the peak strength is found to be occurring either at a lower axial strain with increase in relative density at the same confinement pressure, or at a higher axial strain with increase in confinement pressure at the same relative density. The strength at higher strain level comes closer to each other irrespective of the initial specimen density.

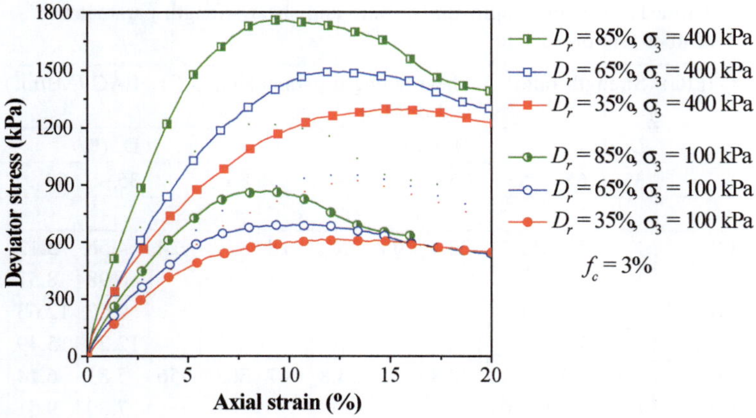

Fig. 4. Effect of relative density and confinement pressure on stress-strain behavior of sand

3.2 Strength Improvement Ratio

Strength improvement ratio is being defined as the ratio of peak deviator stress of reinforced sand to that of unreinforced sand for similar conditions. Table 1 summarizes the strength improvement ratio of all the soil-fiber mixes tested. The strength ratio is found to be maximum at low confinement pressure (σ_3 = 100 kPa) and then tends to decrease generally with confinement pressure at all soil relative densities. This indicates that the effectiveness of fiber-reinforcement is greater at low confining pressure and decreases with increasing confining pressure at any relative density of specimen.

The improvement in strength ratio is found out to be higher for specimens compacted at 35% relative density for all fiber combinations with maximum strength ratio of 1.80 obtained for 3% fiber content. The variation in strength ratio is found to be less for low fiber content (f_c = 0.5%) and gets further pronounced with increase in fiber content at all compacted states of sand.

3.3 Strength Envelopes

Shear strength parameters have been obtained by plotting Mohr-Coulomb envelope. Figure 5 presents a typical Mohr-Coulomb envelope showing the effect of fiber content on the shear strength of dense sand (D_r = 85%). The slopes of the strength envelopes are almost parallel to each other, whereas the intercept of envelope is found to increase with increasing fiber content up to 3% and then decreases for 4% fiber content. The shear strength parameters are summarized in Table 1 for all sand-fiber combinations.

For any relative density, the shear strength is found to increase with fiber content. It can be noted that the improvement in shear strength is mainly due to cohesion induced by fiber reinforcement. This cohesion increases with relative density. However, the friction angle of mixes is found to vary insignificantly with fiber content. The friction angle of sand specimen with 35% relative density initially increases marginally with fiber addition up to 2% fiber content and beyond this it decreases. Whereas for

Table 1. Strength improvement ratio and shear strength parameters of fiber-reinforced sand

f_c (%)	σ_3 (kPa)	Strength ratio			Shear strength parameters, φ (°) and c (kPa)						EAC (MJ/m^3)		
		D_r (%)			D_r (%)						D_r (%)		
		35	65	85	35		65		85		35	65	85
					φ	c	φ	c	φ	c			
0	100	1	1	1	32.7	11	35	18	37	24	4.56	5.12	5.88
	200	1	1	1							6.98	8.57	9.72
	300	1	1	1							9.57	12.63	13.99
	400	1	1	1							12.21	15.49	18.08
0.5	100	1.23	1.16	1.21	32.8	42	34.8	47	36.9	56	5.81	6.14	6.78
	200	1.16	1.10	1.07							7.89	9.51	10.50
	300	1.10	1.08	1.04							10.39	13.29	14.42
	400	1.07	1.06	1.02							12.10	15.99	18.34
1	100	1.41	1.35	1.38	32.9	67	34.6	70	36.8	77	6.63	6.90	7.83
	200	1.34	1.24	1.21							8.77	10.18	11.59
	300	1.19	1.17	1.14							11.17	13.79	15.59
	400	1.16	1.13	1.08							13.39	16.43	19.20
2	100	1.68	1.56	1.66	33.0	89	34.4	96	36.8	110	7.58	7.86	9.50
	200	1.55	1.36	1.35							9.65	11.13	12.95
	300	1.34	1.22	1.24							11.88	14.47	16.40
	400	1.22	1.18	1.17							14.15	16.74	19.82
3	100	1.80	1.73	1.90	32.6	106	34.2	117	36.7	138	8.07	8.56	10.71
	200	1.66	1.43	1.56							10.02	11.72	14.25
	300	1.40	1.31	1.36							12.29	14.98	18.21
	400	1.25	1.26	1.27							14.56	17.03	21.21
4	100	1.76	1.67	1.86	32.4	102	34.0	110	36.7	135	7.64	7.89	10.09
	200	1.58	1.39	1.49							9.68	10.84	13.69
	300	1.38	1.29	1.35							11.93	14.40	17.64
	400	1.23	1.24	1.25							14.00	16.64	21.13

specimens of 65 and 85% relative density, friction angle marginally decreases as fiber content increases. Improvement of cohesion is due to the fact that at the time of loading, presence of fibers reduces the movement of soil particles by sharing the stress coming on them and also provides some extra internal confinement to the specimen. The sharing of load by fibers provides increased tensile strength to the sand matrix and the extra confinement effect of fiber increases the specimen load carrying capacity in terms of increased shear strength.

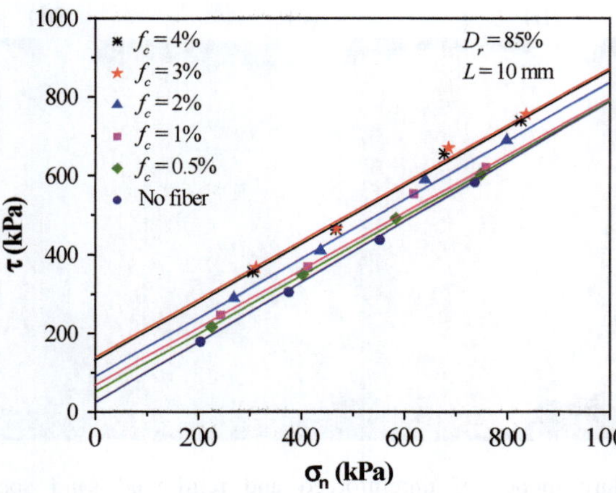

Fig. 5. Effect of fiber content on Mohr-Coulomb envelope of sand (D_r = 85%)

3.4 Energy Absorption Capacity

A relative measure of the improvement in toughness due to fiber inclusion can be obtained by comparing the energy absorption capacity (EAC) of fiber-reinforced sand with the unreinforced soil. The EAC can be obtained by calculating area under stress-strain plot. The EAC values summarized in Table 1 have been calculated by taking into consideration the area under the stress-strain curves up to axial strain of 15%.

The increase in the energy absorption capacity is noticeable with fiber reinforcement for any particular compacted relative density. Due to random distribution of fibers, the progressive absorption of energy takes place during loading. Improved EAC also reflects the improvement in stress-strain and post-peak behavior of reinforced sand. The EAC is further found to increase with increasing density of specimen. Increase in EAC indicates that the reinforced sand requires higher energy to deform and has greater bearing capacity.

3.5 Failure Patterns

Typical failure modes for unreinforced and fiber-reinforced sand are presented in Fig. 6 for different compacted states at 3% fiber content. The unreinforced sand specimen with 35% relative density has undergone bulging failure (Fig. 6a). In contrast, the fiber-reinforced sand specimen of 35% relative density is observed to fail without bulging as the fibers have restrained the lateral spreading of the soil mass (Fig. 6b). The unreinforced sand specimen with 85% relative density has developed clear shear failure plane (Fig. 6c) passing from top to bottom of specimen. However, the shear failure of reinforced sand specimen of 85% relative density has been arrested by the fibers inclusion. Smaller shear plane can be seen in the middle of specimen with some bulging indicating strain localization in a smaller zone (Fig. 6d) with fiber inclusion. It can further be noted that the upper half section of reinforced dense sand specimen

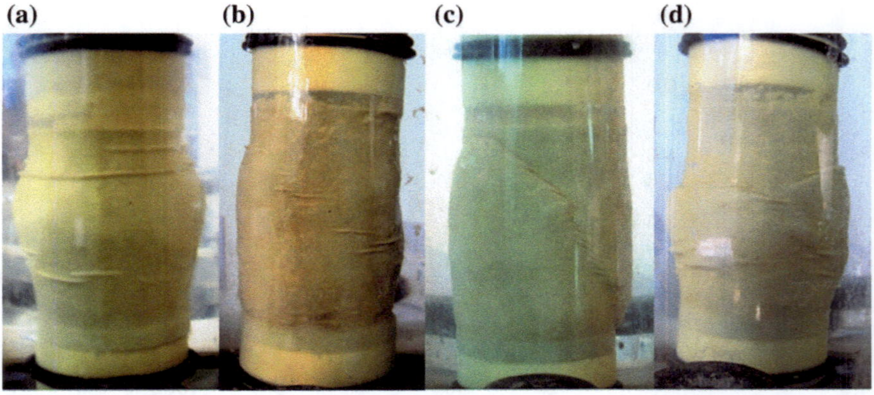

Fig. 6. Typical failure modes of unreinforced and reinforced sand specimens (f_c = 3%): **a** unreinforced with 35% density, **b** reinforced with 35% density, **c** unreinforced with 85% density and **d** reinforced with 85% density

remains like a column and seems to penetrate in lower half section under loading. This indicates that the fiber reinforcement contains the specimen failure in a relatively smaller region.

4 Conclusions

With glass fiber reinforcement, the shear strength or the load carrying capacity of the sand is found to have improved significantly. The post-failure residual strength of the fiber-reinforced sand is also higher than that of the unreinforced sand. As the failure strain increases, it can be concluded that through introduction of glass fibers, any chance of sudden failure of the reinforced sand is significantly reduced. The sand with 3% optimum fiber content can be used for pavement subgrade and foundation beds. For improving the specimen stiffness, it is better to increase specimen density. Inclusion of fiber contains the specimen failure in relatively small zone by restricting the bulging of sand specimen with 35% relative density and shear failure of dense sand specimen with 85% relative density.

References

ASTM D4767-11: Standard Test Method for Consolidated Undrained Triaxial Compression Test for Cohesive Soils, ASTM International, West Conshohocken, PA (2011)

Benson, C.H., Khire, M.V.: Reinforcing sand with strips of reclaimed high-density polyethylene. J. Geotech. Eng. ASCE **120**(5), 838–855 (1994)

Claria, J.J., Vettorelo, P.V.: Mechanical behavior of loose sand reinforced with synthetic fibers. Soil Mech. Found. Eng. **53**(1), 12–18 (2016). https://doi.org/10.1007/s11204-016-9357-9

Consoli, N.C., Festugato, L., Heineck, K.S.: Strain-hardening behavior of fiber-reinforced sand in view of filament geometry. Geosynth. Int. **16**(2), 109–115 (2009). https://doi.org/10.1680/gein.2009.16.2.109

Eldesouky, H., Morsy, M.M., Mansour, M.F.: Fiber-reinforced sand strength and dilation characteristics. Ain Shams Eng. J. **7**(2), 517–526 (2016). https://doi.org/10.1016/j.asej.2015.06.003

Falorca, I.M.C.F.G., Pinto, M.I.M.: Effect of short randomly distributed polypropylene microfibers on shear strength behavior of soils. Geosynth. Int. **18**(1), 2–11 (2011). https://doi.org/10.1680/gein.2011.18.1.2

Hamidi, A., Hooresfand, M.: Effect of fiber reinforcement on triaxial shear behavior of cement treated sand. Geotext. Geomembr. **36**, 1–9 (2013). https://doi.org/10.1016/j.geotexmem.2012.10.005

Li, C., Zornberg, J.G.: Mobilization of reinforcement force in fiber-reinforced soil. J. Geotech. Geoenviron. Eng. ASCE **139**(1), 107–115 (2013). https://doi.org/10.1061/(ASCE)GT.1943-5606.0000745

Li, J., Tang, C., Wang, D., Pei, X., Shi, B.: Effect of discrete fiber reinforcement on soil tensile strength. J. Rock. Mech. Geotech. Eng. **6**(2), 133–137 (2014). https://doi.org/10.1016/j.jrmge.2014.01.003

Lirer, S., Flora, A., Consoli, N.C.: On the strength of fiber-reinforced soils. Soils Found. **51**(4), 601–609 (2011). https://doi.org/10.3208/sandf.51.601

Michalowski, R.L., Cermak, J.: Triaxial compression of sand reinforced with fibers. J. Geotech. Geoenviron. Eng. ASCE **129**(2), 125–136 (2003). https://doi.org/10.1061/(ASCE)1090-0241(2003)129:2(125)

Pino, L.F.M., Baudet, B.A.: The effect of particle size distribution on the mechanics of fiber-reinforced sand under one-dimensional compression. Geotext. Geomembr. **43**(3), 250–258 (2015). https://doi.org/10.1016/j.geotexmem.2015.02.004

Tang, C.S., Shi, B., Zhao, L.Z.: Interfacial shear strength of fiber-reinforced soil. Geotext. Geomembr. **28**(1), 54–62 (2010). https://doi.org/10.1016/j.geotexmem.2009.10.001

Tingle, S., Webster, S.L. Santoni, R.L. Discrete fiber reinforcement of sands for expedient road construction. Technical Report GL-99-3, U.S. Army Corps of Engineers Waterways Experiment Station (1999)

Yetimoglu, T., Salbas, O.: A study on shear strength of sands reinforced with randomly distributed discrete fibers. Geotext. Geomembr. **21**(2), 103–110 (2002). https://doi.org/10.1016/S0266-1144(03)00003-7

Comparative Analysis on the Effect of Asphalt Film Aging Test and Actual Production, Transportation and Paving on Asphalt Aging Degree

Qingqing Zhang(✉) and Zhichao Pu

Cangzhou Municipal Engineering Company Limited, Cangzhou, China
czszjsk@126.com

Abstract. Degrees of aging on hot-mix asphalt mixture depend on the processes utilized on producing, transporting and paving operations. Rotated thin film oven (RTFOT) and/or thin film oven test (TFOT) tests were commonly used to determine the performance of asphalt aging in the laboratory. In this paper, the samples from the field (through actual mixing, transporting, and placing) were tested and compared with the samples from the lab (through RTFOT simulation). Through the analyses of the test results, it is concluded that the degrees of aging are much higher for asphalt samples from the field than those samples from lab aging simulations (i.e. RTFOT test).

1 Introduction

Asphalt aging can reduce adhesion, water stability performance, and thermal crack resistance of asphalt mixture [1, 2]. During mixing, transporting, and placing, hot mix asphalt mixtures need to be covered and kept at high temperature. Maintaining at high temperature ensures the good workability of hot mix asphalt mixture. Meanwhile, it also causes the complex physical and chemical changes of asphalt mixture that leads to aging [3].

Usually, thin film oven test (TFOT) and rotation thin film oven test (RTFOT) were employed to simulate asphalt aging of construction. However, there are not many studies devoted to understanding asphalt aging in asphalt mixtures in China. Based on ASTM D2127 [4] and D5404 [5], the paper investigated asphalt aging under different construction condition by asphalt separation centrifuge extractor and rotary evaporator. In addition, this paper discussed the relationship between lab aging and actual onsite aging. Methodologies to reduce asphalt aging and improve asphalt pavement durability are proposed in the paper.

© Springer International Publishing AG, part of Springer Nature 2019
A. Sevi et al. (eds.), *Enhancements in Applied Geomechanics, Mining, and Excavation Simulation and Analysis*, Sustainable Civil Infrastructures,
https://doi.org/10.1007/978-3-319-95645-9_15

2 Test Method

To extract aged asphalt from mixture, asphalt separation centrifuge extractor and rotary evaporator were utilized in this study. To investigate the effects of test instruments on aged asphalt test results, basic engineering indexes of origin asphalt, extracted asphalt from asphalt trichloroethylene solution, and extracted asphalt from asphalt mortar which was mixture of origin asphalt and mineral filler have been measured and comparisons were made.

3 Test Instrument

Digital asphalt penetration tester (Type WSY-026), automatic asphalt soft point tester (Type SYD-2806-E), low temperature ductility tester (Type LYY-9B), RTFOT tester (Type SYD-0610), asphalt separation centrifuge extractor (Type 20-1100), and rotary evaporator (Type BUCHI) were employed to measure engineering indexes.

4 Validation of Test Method

The origin asphalt used in the study is 70# (Penetration 60 ~ 80) asphalt. The asphalt solution was prepared by trichloroethylene to dissolve the origin asphalt, and the asphalt was extracted by the rotary evaporator. The mineral filler for origin asphalt weight ratio to asphaltic mortar was 1.6. The dissolved mineral filler in the solution was extracted by asphalt separation centrifuge extractor and rotary evaporator. Test results were shown in Table 1.

Table 1. Recovery of mineral powder

Test item	Results
70# origin asphalt/g	60.82
Mineral filler/g	97.31
Extracted mineral filler from the solution/g	97.06
Weight proportion of extracted filler to initial filler	99.7

The proportion of extracted filler was up to 99.7% from Table 1, which indicated that there was a little mineral filler missed during separation centrifuge and rotary evaporation process.

The basic engineering indexes [6] of asphalt were tested after three treatments, and the results were shown in Table 2.

The results in Table 2 showed that there were some differences in engineering indexes of origin asphalt and the asphalt extracted from trichloroethylene asphalt solution: the penetration of extracted asphalt decreased 0.05 mm, soft point decreased 0.6 °C, and no change in ductility. The asphalt extracted from asphalt mortar was harder than origin asphalt: its penetration increased 0.09 mm, soft point increased 0.3 °C,

Table 2. Basic properties of bitumen after three treatments

Asphalt type	Basic performance index			
	Penetration (25 °C, 5 s, 100 g)/0.1 mm	Soft point (ring and ball method)/°C	Ductility (15 °C)/cm	Ductility (10 °C)/cm
Origin asphalt	76.3	47.3	>100	>100
The asphalt extracted from trichloroethylene asphalt solution by rotary evaporator	75.8	46.7	>100	>100
The asphalt extracted from asphalt mortar by separation centrifuge and rotary evaporation	75.4	47.6	>100	>100

and ductility kept stable. Minor variability in the numerical value met the standard permissible error requirement of repetition, reproducibility, which meant asphalt separation centrifuge extractor and rotary evaporator had no effect on the three basic engineering indexes. The studies of Fujie Zhou [7] and Hongsheng Li [8] also confirmed asphalt separation centrifuge extractor and rotary evaporator effectiveness for study on extracted asphalt.

5 Test Results

Ten different $70^{\#}$ origin asphalt were included in the study, and their engineering indexes were shown in Table 3. RTFOT test was used to simulate short time aging [3, 6], and results were shown in Table 4.

Table 3. Test results of 10 origin asphalts

Test item	No. 1	No. 2	No. 3	No. 4	No. 5	No. 6	No. 7	No. 8	No. 9	No. 10	Average
Penetration (25 °C, 5 s, 100 g)/0.1 mm	69.2	78.2	78.1	74.6	69.2	72.3	74.9	72.5	68.3	65.8	72.3
Soft point (ring and ball)/°C	47.2	46.7	47.7	46.5	46.2	47.1	46.8	47.3	47.6	47.1	47.2
Ductility (15 °C)/cm	>100	>100	>100	>100	>100	>100	>100	>100	>100	>100	>100
Ductility (10 °C)/cm	>100	>100	>100	>100	>100	>100	>100	>100	>100	>100	>100

Table 4. Test results of origin asphalts after RTFOT

Test item	I-1	I-2	I-3	I-4	I-5	I-6	I-7	I-8	I-9	I-10	Average
Penetration (25 °C, 5 s, 100 g)/0.1 mm	46.8	48.3	48.7	48.2	43.3	45.1	46.5	44.6	43.7	44.9	46.0
Soft point (ring and ball)/ °C	52.4	52.0	51.3	54.3	52.9	51.7	53.7	54.9	53.9	52.4	53.0
Ductility (10 °C)/cm	8.4	7.9	6.5	9.3	7.6	7.8	6.4	6.9	6.8	8.7	7.6

Meanwhile, the mixing plant selected these origin asphalts to produce asphalt mixtures with the same aggregate gradation. After placing, aged asphalts have been extracted from mixtures by asphalt separation centrifuge extractor and rotary evaporator. The corresponding engineering indexes are shown in Table 5.

It view of Tables 3 and 5, it is found that extracted asphalt from mixture were quite different in engineering indexes compared with origin asphalt as penetration decreased significantly, soft point increased drastically, and ductility decreased significantly. These showed that extracted asphalt from mixture became stiffer and more brittle that lead to lower thermal-crack resistance. It indicates that extracted asphalts show signs of significant aging.

Table 5. Test results of extracted asphalts from mixtures

Test item	II-1	II-2	II-3	II-4	II-5	II-6	II-7	II-8	II-9	II-10	Average
Penetration (25°C, 5 s, 100 g)/0.1 mm	33.8	17.8	44.2	36.8	34.7	38.6	46.2	18.6	36.7	43.4	35.1
Soft point (ring and ball)/°C	66.6	75.3	55.6	55.5	57.2	53.6	54.9	71.6	57.3	54.1	60.2
Ductility (15 °C)/cm	3	1	9	10	34	23	10	1	8	15	11.4
Ductility (10 °C)/cm	1	<1	3	3	4	3	4	1	3	7	2.4

By comparing Tables 4 and 5, it was found that 10 asphalt samples extracted from mixture had smaller penetration, higher soft point temperature, and lower ductility as compared with sample after RTFOT. It is concluded that mixture samples after RTFOT show signs of significant aging.

After RTFOT, penetration and soft point of asphalt samples extracted from mixtures were compared with those of origin asphalt samples, as shown in Table 6.

In terms of retained penetration ratio in Table 6, the minimum value among samples after RTFOT was 62%, but only two asphalt samples were extracted from mixtures that had higher retained penetration value of more than 62%. The minimum value among samples extracted from mixture was 23%. Most of samples became more consist and stiffer. For softening point, it was found that the maximum and minimum increased values among samples after RTFOT were 17, and 8% respectively. Note that

Table 6. Comparisons of asphalt aging between lab simulation and field samples

Serial number	Retained penetration ratio (%)		Soft point increment (%)	
	Sample after RTFOT/origin sample	Sample extracted from mixture/origin sample	Sample after RTFOT/origin sample	Sample extracted from mixture/origin sample
1#	68	49	11	41
2#	62	23	11	61
3#	62	57	8	17
4#	65	49	17	19
5#	63	50	15	24
6#	62	53	10	14
7#	62	62	15	17
8#	62	26	16	51
9#	64	54	13	20
10#	68	66	11	15
Average	63.8	48.9	12.7	27.9

maximum and minimum increases among samples extracted from mixture were 61, and 14%, respectively. Softening point of samples extracted from mixtures had obvious increased, which mean reducing anti-aging ability, and more negative impact on engineering projects.

In this study, the AC-20 mixture is made of skeleton suspended hot mix asphalt and limestone was used as gravel. Before mixing, $70^{\#}$ asphalt was usually heated to 155 ~ 165 °C, and the temperature of mineral aggregate was higher than asphalt temperature 10 ~ 30 °C for drying and heating. During mixing, asphalt contacted with aggregate particles in thin film state and aged most rapidly. The mixing time was limited in 3 min, but it played important role on aging of asphalt [10]. The average temperature of asphalt mixture is about 160 °C, and the temperature coefficient of variation is below 1.6%, so we think the discharge temperature of the mixture is uniform and stable. The time hot mix asphalt mixture produced by mixing factory, then shipped to the construction site is shown in Table 7. The spread speed of asphalt mixture is controlled at 2 m/min to ensure uniform spread. The average temperature of paved asphalt mixture is 145 °C. Compaction is divided into three stages which are initial compaction, repeat compaction and final compaction. As can be seen from Table 7, the transportation time of the Nos. 2, 8 and 9 samples exceeds 2 h, and the deterioration shown in Table 5 is more serious than that of other samples.

In the process of storage, transporting and placing of hot mix asphalt, the asphalt mix is kept in the high temperature and thus asphalt aging was continuous. Although

Table 7. The transport time of the mixture

Test item	III-1	III-2	III-3	III-4	III-5	III-6	III-7	III-8	III-9	III-10	Average
Transport time/h	1.5	2.5	0.8	1.9	2.0	1.7	0.7	2.3	2.1	1.8	1.7

the rotary evaporator method would cause a little degree of aging during the process of asphalt extraction, the influence is relatively limited as comparing to the origin asphalt [11–13]. So most aging of asphalt extracted from field samples occurred in the stages of mixing, transporting, and placing.

6 Conclusions

It was found that there was low influence on basic engineering index properties of extracted asphalt from mixture by using asphalt separation centrifuge extractor and rotary evaporator.

Through the analyses of the test results, it is concluded that the degrees of aging are much higher for asphalt samples from the field than those samples from lab aging simulation. Because of imitated field samples, it could not represent all field conditions. Thus, it is suggested that additional studies for asphaltic aging and their associated components should be investigated thoroughly.

It was recommended to strictly control the heating time of asphalt and to minimize any delays on mixing, transporting and placing. It is because the higher the heating temperature of asphalt and mineral aggregate, the greater impacts on the aging of asphalt. Thus, it is advisable to strictly control the mixing temperature to a certain temperature interval. In addition, heating and drying temperature of aggregate should be reduced as much as possible and the mixing time should be limited. It further suggests of the benefits of utilizing warm mix techniques.

References

1. Ma, Q., Rong, G., Juan, Z., et al.: Analysis on the influence of asphalt aging on the properties of asphalt and asphalt mixture [J]. Highway traffic technology **3**, 76–78 (2012)
2. Lu J.: Study on aging behavior and recycling technique of asphalt pavement [D]. Chang An University (2008)
3. Shi, Z., Yin, X., Zhang, W., et al.: Effect of asphalt aging methods on asphalt performance [J]. Build. Mat. World **33**(5), 45–47 (2014)
4. ASTM: Standard Practice for Recovery of Asphalt from Solution using the Rotary Evaporator D5404. West Conshohocken, PA (2011)
5. ASTM: Standard Test Methods for Quantitative Extraction of Bitumen from Bituminous Paving Mixtures D2127. West Conshohocken, PA (2011)
6. JTG E20-2011.: The test criterion for highway project pitch and pitch compounds [S]
7. Zhou, F.: RAS binder characterization and blending with virgin binders. Annual Meeting of Transportation Research Board, Washington, DC
8. Li, H.S., Wu, Y.B., Guo, Y.F.: Validation of reclaimed shingles asphalt binder extraction and recovery methods. In: Geo-Hubei 2014 International Conference on Sustainable Infrastructure: Advanced Characterization of Asphalt and Concrete Materials, pp. 17–23. Yichang, Hubei, China, American Society of Civil Engineers (ASCE) (2014)
9. He, H., Chen, H.: Evaluation of two experimental methods for the aging performance of asphalt [J]. West. Explor. Eng. **1**, 249–250 (2006)

10. He, Z., Zumin, O., Nan, L., et al.: Effect of heating temperature and heating time on anti-aging performance of asphalt mixture [J]. J. Chongqing Jiaotong University **31**(2), 239–242 (2014)
11. Tian, X., Mo, Y., Zheng, J.: Effect of extracting reclaiming process on aging degree of asphalt [J]. J. Transp. Eng. **5**(2), 23–26 (2005)
12. Zheng, C., Xin, M., Dan, L., et al.: Contrast test of asphalt reclaimed method for aging degree [J]. Highw. Eng. **32**(6), 165–168 (2007)
13. Wang, H., Wang, F.: Discussion on the effects of asphalt aging and pavement performance [J]. China Water Transp. **08**(01), 96–97 (2008)

Response of Swelling Clays to Superstructure Vertical Loads

M. A. Dafalla[1(✉)], E. Mutaz[2], and M. A. Al-Shamrani[1]

[1] Civil Engineering, King Saud University, Riyadh 11421, Saudi Arabia
{mdafalla, shamrani}@ksu.edu.sa
[2] Swissboring & Company Llc., P.O. Box 2694, Ruwi, P.C. 112, Oman
muzjnd_2000@hotmail.com

Abstract. The laboratory testing method for determination of swelling pressure using oedometer apparatus tends to give very conservative values when one dimensional oedometer tests are utilized. Design based on such obtained values leads to unjustified construction cost that adds up to the problems associated with this specific type of soil. This study is aimed at telling the practice and design engineers to fully understand the figures quoted by laboratories and amend them as appropriate to reflect realistic field conditions. Multi-dimensional tri-axial tests are expected to be more informative but due to time and cost constraint it is proposed to utilize a rather simple 3D free swell test along with the oedometer test in order to rationally predict the potential swelling pressure for specific clay. Vertical and volumetric strain relationships obtained in 3D free swelling tests can be utilized to predict the swelling pressure when compared to the profile of load deformation values obtained in one-dimensional swelling tests. The vertical strain associated with load increment as measured during the oedometer test can give a guide for estimating the actual swelling pressure. 3D swell test results of a local clay obtained from Al-Ghatt area in Saudi Arabia were examined and a correction factor procedure for the swelling pressure prediction is suggested and combined with a 3D swell ratio.

1 Introduction

The estimated annual cost related to the problems associated with the expansive behavior of the soil is billions of dollars globally (Jone and Jones 1987). Furthermore, designing sustainable structure able to resist this behavior is very costly. Therefore, a realistic estimation of the swelling pressure attracts the attention of researchers and practitioners alike in order to obtain realistic figures. In order to assess the uplift movement of the swelling soil, the swelling pressure shall be addressed. Swelling pressure can be defined as the pressure required keeping a soil element at a constant volume (Caterina Di Maio 2001). A proper evaluation of swelling pressure is considered as a vital parameter in designing sustainable structures. According to Prabhakara Rao (2013), the swelling of the soils depends on many factors such as type of clay, clay minerals, moisture content and surcharge weight. At a laboratory scale, the determination of swelling pressure can be carried out through different methodologies which are summarized in Table 1.

© Springer International Publishing AG, part of Springer Nature 2019
A. Sevi et al. (eds.), *Enhancements in Applied Geomechanics, Mining, and Excavation Simulation and Analysis,* Sustainable Civil Infrastructures,
https://doi.org/10.1007/978-3-319-95645-9_16

Table 1. Methodologies for the determination of swelling pressure

Methodology	Advantage	Disadvantages
Constant volume swell pressure method	(a) Experimental procedure versatile and results fairly accurate (b) Standardized equipment is readily available in the market (c) Rational estimation of load which could be applied such that the heave developed is tolerable	a) Experimental procedure does not simulate field condition as the weight of the structure in service does not change with time (b) This procedure necessitates uninterrupted presence of the personnel throughout the experiment (c) Experiment consumes relatively longer time
Different pressures test	(a) Soil sample in consolidometer is subjected to uniform pressure all through its thickness (b) Suitable to big projects	(a) Experimental procedure does not simulate field condition as the weight of the structure in service does not change with time (b) This procedure necessitates uninterrupted presence of the personnel throughout the experiment (c) Experiment consumes relatively longer time
Double Oedometer method	(a) Swelling pressure of higher order is produced	(a) It does not require normal sequence of loading-submersion of soil sample
Swell consolidation method	(a) Most favorable to pre-consolidated clays	(a) The pressure required to compress the pre-wetted sample is higher than other methods

After Prabhakara Rao (2013)

In Table 1 the constant volume swell pressure test is conducted with water added frequently to a swelling soil with increments of applied pressure to maintain constant volume. In the different pressure test samples are subjected to a series of loading and unloading until reducing the sample to its original volume. The double oedometer method includes two samples subjected to oedometer tests. One tested for swell under wetting and the second sample tested at the natural water content. In the swell consolidation method both swelling and consolidation tests are carried out together. From logical point of view, if an element is subjected to lateral confinement then its chance to expand vertically is much higher than a freely swelling matrix in 3D. Hence, the major drawback of the conventional oedometer method is the constraint condition of samples with the ability of upward movement only. In order to simulate the actual field condition more closely where a soil element can swell freely in all directions, the 3D-Swell apparatus have been introduced by numerous researchers (Aravind Pedarla et al. 2016;

Mutaz et al. 2011a, b and Al-Shamrani et al. 2010) to evaluate the vertical swelling as well as the volumetric strains. The 3D-Swell setup is comprised of 70 mm compaction mold with 140 mm height. A vertical dial gauge was placed at the top and used to determine vertical elongation of the specimen at select time periods and a pie tape was used to measure lateral expansion of the specimen at same time periods (Al-Shamrani et al. 2010). Once a sample is prepared under certain water content and density conditions, then it will be covered by a membrane and placed inside the bottom of the mold while porous stones were placed at the top and bottom of the tested sample. Before running the test, the vertical dial gauge reading as well as the initial diameter of the sample were recorded and noted as reference readings. Then, water is introduced to the sample from top and bottom with vertical and volumetric expansion monitored.

This study is aimed at providing a guide for the practice and design engineers to help them adjust the figures quoted by laboratories or literature (Dafalla and Al-Shamrani 2012) and amends them as appropriate to reflect realistic field conditions. A research study was conducted on samples from Al-Ghatt province in Saudi Arabia. A series of 3D-Swell tests were carried out at untreated and treated samples with different percentages of lime and cement at different conditions of moisture content for the sake of measuring the vertical and volumetric strains and the ratio between these strains. In addition, oedometer tests were conducted in order to measure the vertical swelling as well as swelling pressure at lateral restraint condition. A comparison was done between the 3D-Swell and oedometer test results and a realistic design criterion was presented.

2 Experimental Program

Materials for this study were collected from Al-Ghatt province located 270 km Northwest of Riyadh in Saudi Arabia. The severity nature of Al-Ghatt soil has been addressed by several researchers (Dafalla and Al-Shamrani 2008; Mutaz et al. 2011a, b). Referring to Al-Shamrani et al. (2010), Al-Ghatt soil has been characterized as high expansive clay with presence of expansive minerals pertaining to smectite group. Basic soil tests were carried out for Al-Ghatt clay such as; gradation, hydrometer, Atterberg limits, specific gravity and compaction test. According to the gradation test, the percentage of fine material is 86%. From Atterberg limits, the liquid limit, plastic limit and plasticity index are 60, 30 and 30 respectively. The specific gravity value is 2.8. Al-Ghatt clayey soil is classified as CH in accordance with the unified soil classification system (USCS). Compaction tests were conducted in order to determine the optimum moisture content and its relevant maximum dry density. In order to study the relationship between vertical and volumetric strain for virgin and treated clays, Al-Ghat clay was stabilized with different percentage of lime (4 and 8%) and cement (3 and 6%) at different conditions of moisture; namely OMC (optimum moisture content), DOMC (dry of optimum moisture content) and WOMC (wet of optimum moisture content). Compaction curves of untreated and treated clays are depicted in Fig. 1.

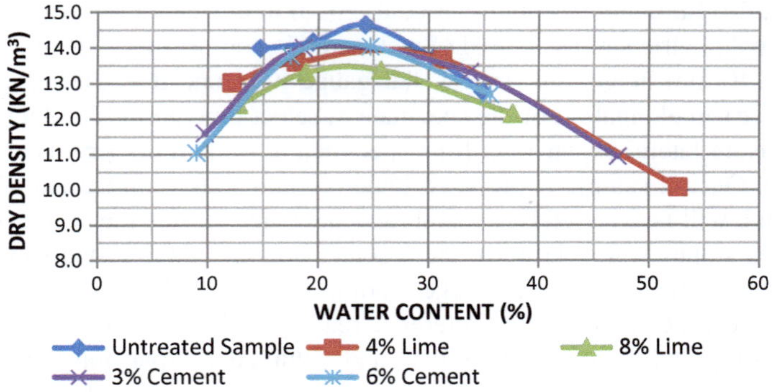

Fig. 1. Compaction curves for untreated and treated Al-Ghatt clay

2.1 3D-Swell Test

In order to predict the swelling percentage under unconfined condition, 3D-swell test was carried out as suggested by Al-Shamrani et al. (2010). Both vertical and volumetric swell strain were measured under a low seating pressure (less than 1 kPa) by using vertical dial gauges and pie tape; respectively. The 3D-swell apparatus is composed of 70 mm diameter compaction mold with 140 mm height. Accordingly, all samples were prepared under different moisture contents and densities and enclosed in a membrane. Free water path; vertically and laterally was assured along the tested samples. The 3D-Swell tests were carried out for different Al-Ghatt specimens prepared at different moisture conditions (OMC, WOMC, and DOMC) with different chemical agents (4, 8% lime and 3, 6% cement). WOMC and DOMC conditions were prepared at dry densities relevant to 95% of the maximum dry density obtained from the compaction curve. All tests have been carried out until the vertical and volumetric readings stabilized for a minimum of 7 h. The results of vertical and volumetric strains for all samples as well as the ratio between both strains are presented in Table 1. It is to be noted that, the average ratio of vertical and volumetric strain results is 0.32. This ratio varies from 0.53 to 0.22. The higher values are associated with untreated samples.

2.2 Oedometer Test

This test normally provides the swelling percentage and swelling pressure under confined ring condition. In this setup, the clayey samples were prepared at a density relevant to the density of WOMC condition from the compaction curve. The samples were thoroughly mixed and prepared in 2.5 inch diameter rings with 16 mm height and subjected to a 7 kPa as a seating load. Then, the whole matrix was submerged with water and allowed to swell vertically. After the vertical reading was stabilized, the difference between the initial and final reading was used to obtain the swelling percentage. The applied pressure required to bring the swell to the original height was recorded as the swelling pressure. The maximum recorded swelling percentage is 9.5% for untreated sample at WOMC condition. The vertical swelling in mm for Al-Ghatt

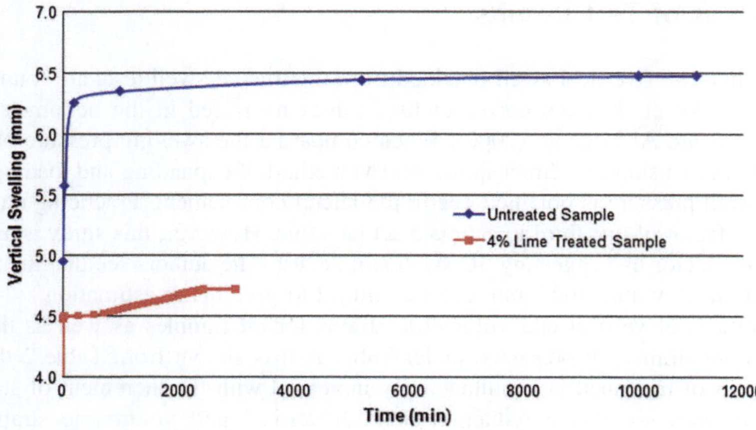

Fig. 2. Influence of lime addition on the vertical swell in an oedometer test for Al-Ghatt clay

untreated clay is reduced by adding lime. Figure 2 shows a graph of typical reduction in vertical swell of Al-Ghatt clay by 4% lime addition. The swelling pressure required to bring the vertical swell to the original height is assumed at three stress levels. Stress level 1 is for untreated Al-Ghatt clay and stress level 2 is for Al-Ghatt clay treated with 4% lime and stress level 3 is for Al-Ghatt clay treated with 8% lime. It can be demonstrated that stress level 1 is greater than stress level 2 and stress level 2 is greater than stress level 3. No quantitative figures are given in this stage of the research study. Figure 3 presents the axial/volumetric strain ratio as obtained in a 3D free swell test for the three stress levels defined for Al-Ghatt clay.

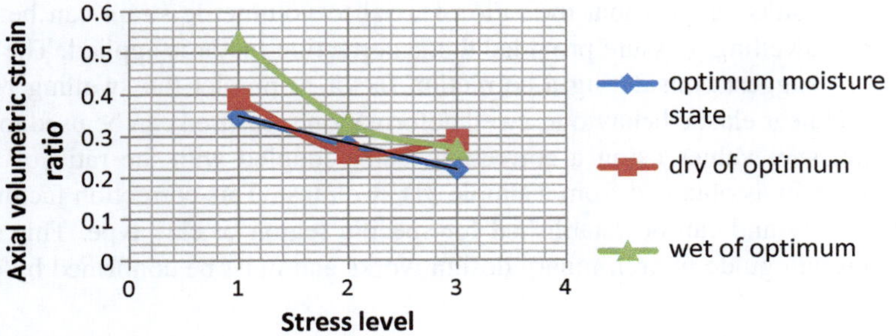

Fig. 3. Presents the axial/volumetric strain ratio as obtained in a 3D free swell test for the three stress levels defined for Al-Ghatt clay

3 Analysis of Test Results

The percentages of vertical swell obtained from the triaxial swell tests are found to be considerably lower than the corresponding values measured in the oedometer tests. Al-Shamrani and Al-Mhaidib (2000). When comparing the swelling pressure obtained in an oedometer using one dimensional ASTM method of expanding and loading, none realistic swell pressure is obtained due to the lateral confinement. Practicing engineers often use a factor of one third to estimate actual value. However, this study is aimed at refining this factor by suggesting 3D swell simple test. The authors see that the ratio of vertical strain to volumetric strain can be utilized to give better estimation.

The results of vertical and volumetric strains for all samples as well as the ratio between both strains are presented under Table 2. It is shown from Table 2 that, the effectiveness of treatment is simultaneously increased with the increment of stabilizer dose. Moreover, significant reduction in both vertical and volumetric strains was observed with the treatment for DOMC and WOMC conditions. Therefore, stabilizer dose and compaction condition play a vital role in the effectiveness of the treatment. This is similarly valid for cement treated clays but only lime was used to demonstrate the suggested approach for this procedure. The data for clay stabilized with cement indicates a similar trend of that shown for lime stabilized clay. Al-Mhaidib (1998) suggested that triaxial tests provides lower values of vertical swell than swell measured from one dimensional oedometer tests. His work indicated that for all swelling pressure values, the ratio of triaxial swell to the oedometer swell is constant and estimated at 0.333. This is not likely to be the case when the mineralogy of the clay or clay content is variable. The current study suggests that the axial to volumetric strain ratio decreases with decrease of the swell pressure. Figure 2 does not suggest quantitative values but the decrease with decrease is clearly shown. The constant volume method and the one dimensional oedometer method were compared for identical sample and results obtained showed that the oedometer method gives higher swelling pressure values than the constant volume method, El Fatih and Muawia (1984).

The test results suggest that the ratio of axial to volumetric strain can be used in estimation of swelling pressure provided that a correction factor is applied. The authors of this work suggest considering a correction factor to predict the swelling pressure. Assuming a linear elastic behaviour, swell deformation measured can be used to correct swelling pressure values using a correction factor coupled with the ratio of axial to volumetric strain as obtained from a simple 3D swell test. This correction factor can be of local validity and can be established for specific region or clay type. This shall be taken as a rough guide in preliminary design works and must be confirmed by detailed testing.

Table 2. 3D-swell test strain results

Sample condition	Vertical strain (%)	Volumetric strain (%)	Ratio of vertical/volumetric strain	% of improvement in vertical strain
Untreated sample at OMC condition	9.15	25.498	0.35	–
Untreated sample at DOMC condition	12.61	32.24	0.39	–
Untreated sample at WOMC condition	9.07	17.11	0.53	–
4% Lime sample at OMC condition	8.35	29.73	0.28	8.74
4% Lime sample at DOMC condition	6.14	23.24	0.26	51.31
4% Lime sample at WOMC condition	6.59	19.74	0.33	27.30
8% Lime sample at OMC condition	5.63	25.17	0.22	38.47
8% Lime sample at DOMC condition	7.71	25.74	0.29	38.86
8% Lime sample at WOMC condition	4.94	18.23	0.27	45.53
3% Cement sample at OMC condition	6.91	19.40	0.35	24.48
3% Cement sample at DOMC condition	6.67	15.13	0.44	47.11
3% Cement sample at WOMC condition	4.21	14.63	0.28	53.58
6% Cement sample at OMC condition	5.97	19.82	0.30	34.75
6% Cement sample at DOMC condition	5.59	26.00	0.21	55.67
6% Cement sample at WOMC condition	3.31	12.44	0.26	63.50

Actual swelling pressure = Oedometer swelling pressure × ratio of axial to volumetric strain × C. C is a correction factor obtained as a correlation from a series of samples belonging to particular area.

Relationship between free axial swell and swelling pressure vary for different soils and different areas but as an example we can present the work of Kaybali and Demir (2011) shown in Fig. 4 of this paper.

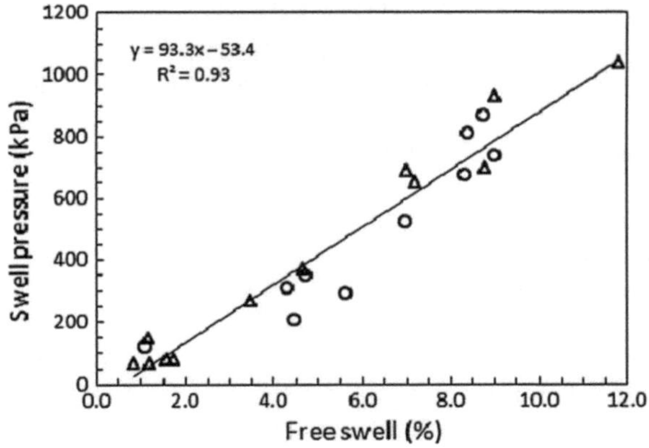

Fig. 4. Relationship between swell pressure and free swell. After Kaybali and Demir (2011)

4 Conclusion

The ratio of axial to volumetric strain obtained in a 3D swell test can be useful in predicting actual swelling pressure when only one dimensional oedometer testing is performed. Expanded and loaded test method tends to overestimate the swelling pressure due to the confinement of sample during the test. It is suggested to use the 3D swell test as an additional tool to better predict the swelling pressure obtained from one dimensional oedometer test. This study suggests an equation for predicting the actual swelling pressure. The actual swelling pressure can be obtained by multiplying the oedometer swelling pressure by the 3D swell test ratio of axial to volumetric strain and a regional correction factor. Regional and area correction factor can be obtained from testing a series of samples.

References

Al-Mhaidib, A.: Prediction of swelling potential of an expansive shale. In: Proceedings of the Second International Conference on Unsaturated Soils, 27–30 Aug 1998, Beijing, China (1998)

Al-Shamrani, M., Al-Mhaidib, A.: Swelling behavior under oedometric and triaxial loading conditions. Geo-Denver 2000, 5–8 Aug 2000, Denver, Colorado, United States (2000)

Caterina Di Maio: Swelling pressure of clayey soils: the influence of stress state and pore liquid composition. Rivista Italiana di Geotecnica 3/2001 (2001)

Dafalla, M.A., Al-Shamrani, M.A.: Expansive soil properties in a semiarid region. Res. J. Environ. Earth Sci. **4**, 930–938 (2012)

Dafalla, M.A., Al-Shamrani, M.A.: Performance-based solutions for foundations on expansive soils-Al-Ghatt region, Saudi Arabia. In: Geo-Chiangmai 2008, Chiangmai, Thailand (2008)

El Fatih, M.A., Muawia, A.D.E.: Comparison of two methods for the measurement of swelling pressure. In: Proceedings of the Fifth International Conference on Expansive Soils, Adelaide, South Australia, 21–23 May 1984

Jone, D.E., Jones, K.A.: Treating expansive soils. Civil Eng. ASCE **43**, 11 (1987)

Kayabali, K., Demir, S.: Measurement of swelling pressure: direct method versus indirect methods. Can. Geotech. J. **48**(3), 354–364 (2011)

Mutaz, E., Dafalla, M.A., Shamrani, M.A.: Stabilization of Al-Ghatt clay shale by using a mixture of lime and cement. In: Geo-Hunan Conference, China, American Society of Civil Engineers—Geotechnical Special Publication GSP 223 (2011a)

Mutaz, E., Shamrani, M.A, Puppala, A.J., Dafalla, M.: Evaluation of chemical stabilization of a highly expansive clayey soil. Transp. Res. Board J. (Washington DC, USA, Low-Volume Roads 2011) **2**, 148–157 (2011b)

Pedarla, A., Puppala, A.J., Laureano, R.H.: Evaluation of swell behavior of expansive clays from internal specific surface and pore size distribution. J. Geotech. Geoenviron. Eng. **142**(2) (2016)

Prabhakara Rao, V.V.N.: Swelling pressure of soil using a predictive tool. Int. J. Comput. Appl. (0975–8887) **63**(19) (2013)

Shamrani, M.A, Mutaz, E., Puppala, A.J., Dafalla, M.A.: Characterization of problematic expansive soils from mineralogical and swell characterization studies. American Society of Civil Engineers—Geotechnical Special Publication GSP 199, Geoflorida, Florida, No. 199, pp. 793–802 (2010)

Author Index

© Springer International Publishing AG, part of Springer Nature 2019
A. Sevi et al. (eds.), *Enhancements in Applied Geomechanics, Mining, and Excavation Simulation and Analysis,* Sustainable Civil Infrastructures,
https://doi.org/10.1007/978-3-319-95645-9

MIX
Papier aus verantwortungsvollen Quellen
Paper from responsible sources
FSC® C105338

If you have any concerns about our products,
you can contact us on
ProductSafety@springernature.com

In case Publisher is established outside the EU,
the EU authorized representative is:
Springer Nature Customer Service Center GmbH
Europaplatz 3, 69115 Heidelberg, Germany

Printed by Libri Plureos GmbH
in Hamburg, Germany